AF334988

INTERNATIONAL ENCYCLOPEDIA *of* UNIFIED SCIENCE

Foundations of Physics

By

Philipp Frank

VOLUMES I AND II · FOUNDATIONS OF THE UNITY OF SCIENCE

VOLUME I · NUMBER 7

International Encyclopedia of Unified Science

Editor-in-Chief   Otto Neurath†
Associate Editors   Rudolf Carnap   Charles Morris

Foundations of the Unity of Science

(Volumes I–II of the Encyclopedia)

Committee of Organization

Rudolf Carnap   Charles W. Morris
Philipp Frank†   Otto Neurath†
Joergen Joergensen   Louis Rougier

†Deceased.

The University of Chicago Press, Chicago 60637
The University of Chicago Press, Ltd., London

Contents:

Contents

I. Introduction

In the present monograph an attempt is made to present physics in such a way that it is fit to become part of unified science. The first step is to bring it into such a shape that it becomes clear which statements tell something about observable facts and which are statements about the choice of symbols. This means that we must discuss the operational meaning of all symbols used and the kind of relations which exists between these symbols.

We are not going to set up a complete system of symbols and operational definitions from which one could derive all facts of physics. Such a systematic presentation would be a very hard job, and at that I suspect that only a very few scientists would read it. Most of this presentation would be a dull routine work, and the reader would not know how to find the points to which he should direct his attention. The scientist is interested in logical analysis if, and only if, this analysis is not trivial or commonplace. One does not need to occupy each square foot of a territory in order to control it. What matters is only the control of some key positions. In this presentation of physics we shall attempt to bring under control only a few key positions in the huge symbolic structure by which the scientists have been "mapping" the wide domain of physical phenomena.

It is not difficult to find out these key positions. We must look only for the starting-points of current "interpretations" of physics on behalf of some philosophical creed, usually idealistic, or even spiritualistic. If we examine these "interpretations", we notice almost regularly that they are not interpretations of physics itself but interpretations of the symbolic structure, ignoring the operational meaning of these symbols. This means that they ignore the link between symbols and observable phenomena. Examples of this type of interpretation are obvious. The "energetic" aspect of physics is interpreted as a "refutation of materialism" and the four-dimensional presentation of relativity theory as establishing the "existence" of a four-dimensional world.

It would be a great mistake to say that the philosophers are chiefly responsible for these misinterpretations. As a matter of fact, each one of these "interpretations" has had its origin in a confused presentation of physics given by a physicist. Some of the latter have been honestly surprised at the fruits of their work. They have blamed the confused way of presenting their subject on the present "crisis" of physics, which leaves this field in a state of transition. One physicist suggested even putting up a sign telling all nonphysicists and in particular philosophers: "Keep out, while repairs are going on!" This attitude of "isolationism" would frustrate all attempts toward the unification of science and is, therefore, directly opposed to the aims of this *Encyclopedia* in general and of this monograph in particular.

Physics has been for centuries the spearhead of advance in human thought. In a unified science it should keep its role as a description of the physical universe and should not deteriorate into an incoherent, and somehow mysterious, agglomeration of symbols, rules, and recipes. In this situation quite a few physicists have tried to avoid all these difficulties by "sticking strictly to the facts" and by keeping away from the dangerous enterprise of logical and critical analysis. There is no doubt that this attempt is doomed to failure. Scientists who looked at the world from such different angles as Ernst Mach and A. N. Whitehead have agreed on one point: a physicist who dodges all logical analysis and tries to be a "physicist and only a physicist" will imbue the presentation of his subject with some "chance philosophy", usually a very obsolete one.[1] A careful examination of the presentation of physics given by physicists who pretend to keep strictly to the "facts and only to the facts" will reveal soon that these presentations are written in a spirit of medieval scholasticism. We can notice it even by a perfunctory look at textbooks for beginners. Quite a few make a generous use of words, like "entity", which

are borrowed from medieval philosophy and have no status in science.

Obsolete philosophical views in physics are mostly obsolete physical theories in a state of petrification.[2] Aristotle's philosophy of physics is a petrification of a physical theory which covered the experiences of Greek and oriental artisans about physical phenomena. Kant's *Metaphysical Principles of Natural Science* is a petrification of Newton's physics. Eddington's *Philosophy of Physical Science* is a petrification of Einstein's relativity and Bohr's quantum theory.

Remembering these facts, we can easily understand why a certain type of confusion in the presentation of physics gives rise to philosophical "interpretations". The lack of coherence in the presentation of a new physical theory often has its origin in the incoherence between the languages of old and new physical theories. The old language slips easily into the presentations of recent physics. For it pretends to express a certain philosophy which is based upon intuition or good sense and cannot be disregarded by physicists. Actually this language has been invented in order to present older physical theories. Its use means, therefore, confusing the language of recent physics with the language of older and abandoned theories. We can now understand why the key points in the structure of physics are the points where philosophical misinterpretations slip in easily. A coherent presentation of these key points and their vicinity will lead easily to a coherent presentation of the whole domain of physics.

The elucidation of the relations between symbols and facts in physics owes much to the works of men like Ernst Mach, Henri Poincaré, C. S. Peirce, and P. W. Bridgman. The schools of thought which have been called Positivism, Pragmatism, and Operationalism have done a fine job. The present monograph follows their lines in general but does not make any attempt to build up three "self-consistent" systems of Positivism, Pragmatism, and Operationalism. The spirit of all these schools of thought is one and the same. But the theses which we would obtain if we tried to make a clear-cut distinction between them would result in three new branches of metaphysics: positivistic, pragmatistic, and operationalistic metaphysics.[3]

II. The Logical Structure of Physical Theories

1. Equations, Logical and Semantical Rules

Every physical theory consists of three essential parts.[4] There are, first, the equations of the theory, e.g., in mechanics, Newton's equation of motion; in electromagnetism, Maxwell's equations of the electromagnetic field; etc. These equations contain terms like 'co-ordinate', 'time', 'force', 'magnetic field intensity', 'electric conductibility', etc. By themselves they cannot be checked to see whether they are in agreement with the physical facts or not. Nor can we check their logical consistency. Carnap calls these equations the "calculus" of the field of physics.[5] Newton's equations of motion are, e.g., the calculus of mechanics. These equations must have as their basis, as Carnap expresses it, a second calculus by which we can learn what transformation of our equations are allowed without altering their meaning. Mechanics needs as a basis the calculus of algebra and geometry. Only by the application of this "second" calculus can the consistency of the first one be checked.

However, the system of both equations and logical rules (e.g., Newton's equations of motion plus the laws of algebra and geometry) is by no means a system of physical laws. We have to add, as a third part, statements which define the physical meaning of words like 'distance', 'time', 'moment', 'simultaneously', 'force', 'mass', etc. The statements which give these definitions are called by Carnap "semantical rules".[6] Besides the terms of the first calculus ('force', 'mass', etc.) and the terms of the second part, the logical rules ('plus'/'minus', 'and'/'or', etc.), the third part, the semantical rules, contains words like 'iron cube', 'wooden bar', 'warm water', 'one inch long', etc.

These latter words are to be understood in the sense in which they are used in our everyday language. Without this loan taken from the language of our daily life, the language of science would not be shareable knowledge. The semantical rules connect our equations or our calculus with the words of the language of our daily life, at least in physics and, perhaps, elsewhere too. But, in order to avoid any ambiguity, we have to be sure never to apply these words in a wider domain than in the range of their applications in our daily life. In this case this language is understood without controversy by everybody. But it is, e.g., illegitimate to apply the expression 'iron cube' to a body of a size of one million miles through, for we do not even know whether the existence of an iron cube of these dimensions would not violate the laws of physics. The same precaution has to be taken in using words like 'distance' or 'simultaneous'. As Carnap points out, only by adding these semantical rules to the equations do the latter become physical laws which can be checked by experiments.[7]

2. Operational Meaning and Validity of a Theory

In physics these semantical rules consist in the description of physical operations. Bridgman has studied the way the rules have been set up and used by the physicists long before the term 'semantics' had been used in logic.[8] To emphasize their character in physics, Bridgman calls these rules "operational definitions". When we know these rules or definitions, we know "the operational meaning" of a term. We must, therefore, always be aware of the fact that, e.g., Newton's equations of motions by themselves do not contain the theory of motion. Newton's equations with their operational meaning provide us with the theory of motion. There is no way of checking whether the equations of motion by themselves are valid or true or correct. They are only a part of the theory. We can check only whether a system of equations plus the operational definitions is confirmed by a certain experiment or not.

If there is no disagreement with the experiment, we can say only that the system as a whole is "confirmed". But we cannot say that the equations by themselves or the operational definitions by themselves are confirmed. It is possible to alter the equations and the operational definitions in such a way that some conclusions drawn from the whole system remain unaltered.

Speaking exactly, we have to remember that experiments confirm a system which consists not only of two but of three kinds of statements: equations, plus operational definitions, plus logical and mathematical rules. Therefore, it is theoretically also possible to alter the traditional rules of logic without altering the confirmed conclusions.[9] We have only to introduce suitable alterations of the equations and the operational definitions. By an experimental confirmation we cannot demonstrate that the new logical rules are better than the old ones.

We demonstrate only that this new logic is a part of a system of statements which is as a whole in agreement with the experimental facts.

These remarks have their basis in a very simple argument of elementary logic which Bertrand Russell once elucidated by a drastic example: We start from the assumptions that "bread is made of stone" and that "stone is nourishing". Then it follows logically that "bread is nourishing". This statement can be confirmed by experiments. If someone claims for this reason that we have confirmed our assumptions, he would certainly be ridiculous. According to Russell, a great many confirmations of physical hypotheses are of this type.[10]

Now what is the meaning of the assertion that Newton's equations or Maxwell's equations themselves are confirmed by experiments? This assertion means, speaking exactly: It is possible to add to these equations in a fairly simple way such operational definitions that the whole system is in agreement with the observed facts. This confirmation also reveals, to be sure, a property of the equations themselves. If, e.g., Newton's equations had not this property, it would be impossible to find handy operational definitions which, added to these equations, would turn them into confirmed physical statements. We would need operational definitions which are complicated and hard to handle. But if such an addition is possible in a fairly simple way, we would say that Newton's equations themselves are confirmed by experiments. If we speak a little loosely, we may say that in this sense the equations themselves are physical laws.

But, even so, the operational definitions of general terms like 'energy', 'entropy', etc., are of an immense complexity. Bridgman has described the difficulties involved in an elaborate way. According to his analysis, "paper and pencil operations" play a great part. He has pointed out that only under very special conditions is it possible to distinguish by reasonable physical operations even between expressions which have apparently as simple a meaning as 'heat conduction' and 'heat radiation' or 'flux of heat' and 'flux of mechanical energy'.[11]

3. Are the General Laws of Nature Pure Conventions?

There has been a school of thought which has assumed that eventually one could find for any set of equations a set of operational definitions which might be added to these equations and turn them into confirmed physical laws. In this case an experimental confirmation would not reveal any property of the equations at all. The equations by themselves would not say anything about the physical world. However, if operational definitions of some of the symbols in the equations were given, the equations would become operational definitions of the rest of the symbols. Newton's equations of motion, e.g., would no longer be laws of motion. If we added to them the operational definitions of 'acceleration' and 'mass', the equations would become operational definitions of 'force' or a convention how to use the term 'force'.

The school of thought which has ascribed this property to all the fundamental equations of physics has been given the name "conventionalism". The French mathematician Henri Poincaré has been quoted as its most prominent spokesman. Actually, Poincaré emphasized that an experimental confirmation of equations plus operational definitions does not confirm the equations and that, by admitting any imaginable operational definition, we can turn every equation into a confirmed one. This statement can, from the angle of formal logic, hardly be refuted. It is logically equivalent to the statements of section 2. The equations, by themselves, are said to be "valid" or confirmable by experiments only if, by substituting "simple and practical" operational definitions, they become confirmed physical laws. This does not exclude that, by admitting all imaginable operational definitions, almost any system of equations could be converted into confirmed laws, provided that the system is not self-contradictory. If we consistently make the distinction between "simple and practical" operational definitions and arbitrary definitions which may be "complicated and impractical", it becomes clear in what sense the general laws of physics are purely conventional and in what sense they are valid assertions about facts.[12]

We understand this distinction easily if we examine the way in which operational definitions are actually used in physics. A specific operational definition can practically be used only if some specific physical laws are assumed to be valid. What is, e.g., the operational definition of the time distance "one hour"? We may say that this is the time during which the big hand of our pocket watch traverses an angle of 360 degrees. However, we are not interested in the reading of a particular watch. We mean to say that any pocket watch can be used. This means assuming that the hands of all pocket watches proceed with one and the same angular velocity. But this is a statement of a physical law about the behavior of watch springs. Moreover, if we say that a certain phenomenon lasts one hour, we do not mean to say that we can use only a spring watch in order to check the statement. We may as well use a pendulum clock and define one hour as a duration of a certain number of oscillations of this pendulum. But this is only possible if we assume the validity of a physical law: The unwinding of a spring as an effect of its elasticity proceeds at a rate which is proportional to the frequency of the pendulum as an effect of gravity. Since these and similar laws are valid, the expression 'one hour' defined by spring watches or pendulum clocks is helpful in giving a simple description of the actual motions of material bodies. Briefly, an operational definition is "simple and practical" if there are physical laws according to which the numerical result of this definition is identical with the result of other independent operations. Or, in other words, an operational definition is "practical" if it plays an important role in the simplest formulation of valid physical laws.

We must appreciate this fact if we wish to appreciate justly the merits of "conventionalism" in physics and in science in general. Let us consider an example which has been used over and over again for stating the case of conventionalism: the principle of the conservation of energy. We start from the case that only mechanical and heat energy play a role. If we consider an isolated system, we can state: The sum of heat energy H and mechanical energy M remains constant through all interchanges. Mathematically: $H + M =$ Constant. This is certainly no longer true if electromagnetic phenomena also come into play. Then we have to introduce also an electromagnetic energy E, and the principle of conservation of energy would say $H + M + E =$ Constant.

The advocates of conventionalism argue: If the last equation is not confirmed by experiment, we can always add a term U in such a way that the equation $H + M + E + U =$ Constant is confirmed by our experiments. Then this last equation is the operational definition of the term U (unknown kind of energy). However, we note that the electromagnetic energy E has not been defined by our equation of conservation only. There is also a different operational definition of E.

It can be calculated from the electric and magnetic field intensity. Thus we have two operational definitions of E. To say that both these definitions render one and the same result means to assert the validity of a certain physical law. This law gives a practical value to the introduction of the symbol E. The introduction of the unknown energy U would be practical only if we knew a second operational definition of U which is independent of the conservation equation $H + M + E + U =$ Constant. Then the principle of the conservation of energy would no longer be purely conventional but a statement about facts.[13]

4. Misuse of the Vernacular of Physics

In some philosophical discussions the concept of energy has been used in a rather loose way. Speaking of biological or psychological phenomena, quite a few authors have argued as follows: If we examine the phenomena in living organisms, it may be that the sum of the physical and chemical forms of energy remains constant during the living process. If this were the case, it would confirm the hypothesis that life is a physico-chemical phenomenon. Let us now discuss the possibility that the physico-chemical forms of energy would not fulfil the law of the conservation of energy, that, e.g., a living organism would do more work than corresponds to the energy it has absorbed. Even in this case one could uphold the principle that the sum of all forms of energy remains constant. One has only to introduce a new term L (energy of life or vital energy) into the law of conservation. If we denote the sum of all physical and chemical energies by P, we confirm by experiments that P does not remain constant during the phenomena of life. But if we choose L in a suitable way, we can always achieve that $P + L =$ Constant. In this case life could not properly be called a physico-chemical phenomenon. But nonetheless it could be treated by an exact science, a "vitalistic biology", which would contain a law of conservation of energy: The sum of physico-chemical energy and vital energy remains constant.

But in this vitalistic or organismic biology the operational meaning of L (vital energy) is defined by the conservation law $(P + L = $ Constant$)$. This law would state a "convention" as to how to use the word "energy of life L" and would be by no means a law about the physical world. The crucial test of the vitalistic biology is whether we can introduce, besides the state variables of physics and chemistry (mass, electric charge, etc.), a small number of biological state variables from which the energy of life (L) can be computed by a simple formula. Certainly we do not know of any attempt to do so. We have no "practical" operation to define the energy of life.[14]

By comparing the concept of energy in physics with this useless concept of vital energy, we learn to understand the real scientific meaning of the en-

ergy law in physics. We know that in physics the conservation law is not the only operational definition of electric energy; or, in other words, there are other equations in the calculus of physics besides the energy equation which contains the symbol E. In addition to it, the number of state variables upon which the energy law depends is a small one. We shall take up this last argument again in the discussion of the concept of force (sec. 15).

5. Mathematical Description or Physical Explanation?

Frequently a distinction has been made between two types of physical theories. One type starts as we did from a set of equations. This type of theory, we have been told, describes the observable phenomena. It is called, therefore, a phenomenological theory or a mathematically descriptive theory. But there is also, according to the traditional distinction, a second type of theory. Here the physical facts are not only "described" but also "explained". If we treat, e.g., the phenomena of light by a theory of the second type, we present a mechanical model of the ether which transmits light and a mechanical model of the atom which emits light. In this way we have a "causal" theory; we are given the causes of the light phenomena, not only a description. This type of theory is sometimes called a theory which uses pictures instead of equations—a pictorial theory. We have also been told frequently that only this type of theory provides an understanding of nature and gives real satisfaction to the mind of the physicist. We have also been told that only this type of theory is really helpful in finding new facts, while the phenomenological theory is only helpful in recording facts which are already known.

As a matter of fact, the whole distinction between the phenomenological and the causal theory has been much overworked and overrated. If we are given a mechanical model of the ether, we have to formulate the equations of this model. And, possessing these equations, we have to proceed exactly as we did in the mathematically descriptive theory. The only thing which would clearly distinguish the second type of theory would be the requirement that these equations are to be of a particular form. They ought to be derived from the equations of motion, as formulated by Newton to handle the motions in the planetary system. But these equations themselves constitute the phenomenological or "descriptive" theory of a particular domain of physical phenomena, the motions of medium-sized material bodies. Therefore, the requirement that all physical phenomena should be explained by mechanical models would mean assuming that the motion of the smallest particles is covered by the same equations that have been extracted from the motions of medium-sized bodies. This assumption is, of course, far from being self-evident and, for that matter, has turned

out definitely to be an oversimplification. We know very well that the motion of subatomic particles cannot be covered by Newton's laws of motion and therefore by no mechanical model. The statement that the laws of "modern" physics (e.g., relativity theory) are not "causal" laws but only "descriptive" ones has a definite operational meaning only if we use the phrase "causal explanation" in the sense of "derivable from Newton's laws of motion".[15]

6. The Metaphysical Basis of "Explanation"

This definition of "causal explanation" would make sense only if we ascribed to Newton's laws of motion a particular logical status. But doing so would mean a discrimination against any attempt to set up a new foundation of physics. The expression "a physical phenomenon is explained" means in the everyday language of the physicist: "The statements about this phenomenon can be derived from a set of equations and practical operational definitions from which we can also derive the bulk of the physical phenomena which are actually known." To single out a particular type of equations as the only legitimate basis of explanation is not justified by science. All attempts at such a justification have been based upon metaphysics. They have had their roots in the belief that the validity of some equations (e.g., Newton's equations) can be established "philosophically" or "epistemologically" without reference to the observable facts which have to be derived. Otherwise the derivations from Newton's equations would not be more of an "explanation" than the derivation from any other set of statements.[16]

Therefore, the distinction between a description and a causal theory is a purely metaphysical distinction. If we look back into the history of science, we notice that Francis Bacon, the leader of empirical philosophy, accused the Copernican system of being purely descriptive while, according to him, the old Ptolemaic system provided understanding.

Philipp Lenard and quite a few "empirically minded" scientists of the twentieth century accused Einstein's theory of relativity of being purely "descriptive", while Newton's theory was supposed to be causal and explanatory. It is instructive to take note of this similarity to the attitude which traditional philosophers took toward the advance in science. The only apparent difference is a shift in the content of the metaphysical creed. In the sixteenth century a physical phenomenon was said to be "explained" if it could be derived from Aristotle's philosophy, while in the nineteenth century only those phenomena were regarded as "explained" which could be derived from Newton's philosophy. This belief in the exceptional logical status of a particular type of equation had become, by-and-by, so firmly established that its metaphysical foundation was forgotten or driven into the subconscious. But the content of the creed remained alive and was more and more credited to "common sense" or "immediate intuition of nature". The empiricist Francis Bacon did condemn Copernicus, and the physicist Lenard did condemn Einstein, not in the name of metaphysics, but in the name of good sense or "unbiased and strictly empirical description of nature." Even a slight glance at the history of scientific thought shows us that the content of yesterday's metaphysics is today's common sense and tomorrow's nonsense.[17]

7. Pictorial Theories

It has been maintained frequently that the theories using the mechanical models are more "visualizable" or "pictorial" or "intuitive" than the theories starting from equations. I think that this claim has a very weak foundation. Speaking from a purely mathematical point of view, every description by an equation is equivalent to a description by a visualizable diagram or picture. But it seems that the philosophers and scientists who discriminate in favor of pictorial theories mean by this word "similar to a familiar group of sense experiences". As a matter of fact, the mechanical interaction of medium-sized material bodies is a phenomenon which we observe again and again in our daily life. These phenomena are covered by Newton's equations in a very satisfactory way. Every theory which describes new thoughts by comparing them with these familiar phenomena strikes a friendly sounding chord in our minds. Such a theory seems to us intelligible and visualizable for this psychological reason. But to require that every theory should be set up according to this pattern would mean to assume that the laws governing this particular domain of phenomena are sufficient to cover all phenomena of physics, even the motion of the smallest particles with the greatest speeds. If this assumption is once rejected, what would be the point in calling a theory "not visualizable"? It would mean that the theory contradicts statements which have been dropped anyway because of their inadequacy.

A particular difficulty in understanding the meaning of physical theories has arisen from the confusing of a theory which is intuitive or visualizable with a theory which is supposed to be supported by our "inner intuition". Quite a few philosophers have asserted and quite a few scientists have agreed that, e.g., the axioms of geometry can be recognized as true by looking into our own minds. In order to give to this ambiguity the appearance of unambiguity, a particular word has been borrowed from German philosophy and introduced into the English vernacular of philosophy, the ambiguous word "anschaulich", which means "conceivable by our sense observation", but also "conceivable by the efforts of our minds even if not observable at all". These two meanings contradict each other in the application of that word to every actual problem.

III. Classical Newtonian Mechanics

8. Domain of Newtonian Mechanics; Mass

Classical mechanics or Newtonian mechanics is still the cornerstone of all physics. By this theory a group of phenomena is covered around which the research work of science was centered in the period when Galileo started the revolution which gave rise to modern physics. These phenomena are the motions of "medium-sized" material bodies, such as the motion of running cars, of machines, of falling and rolling stones. Newton set up a pattern of description which was adjusted as well to the motion of planets as to the motion of these medium-sized bodies. If we understand Newton's laws of motion which define this pattern, we have made the first step toward understanding physics. A great deal of misunderstanding of the recent physical theories (relativity and quantum theory) is due to the fact that Newtonian physics has been frequently misrepresented. There has been, in particular, a tendency in the teaching of physics to suggest that Newton's laws are almost self-evident. In quite a few textbooks the authors express their bewilderment that such obvious truths have been ignored for such a long time. This kind of presentation is a serious obstacle in the way of a real understanding of physics. One has to emphasize, rather, how great a power of imagination was necessary to set up these laws, how long and strenuous the road has been which has led scientists from the observed facts to Newton's laws. The more self-evident these laws appear to us, the less do we understand them.

For medium-sized material bodies we can tentatively define the "mass" of a body by determining the number of grams which we obtain by reading its weight on a spring-balance at sea-level. This operation is unambiguous, as the weight does not change if we move the balance to any point at sea-level. We note here again that the applicability of an operational definition is based on the physical fact that several operations render an identical numerical result. In the domain of the phenomena which are covered by Newton's laws of motion the following statement holds:

If we impart to a body of a given mass in a well-defined environment a certain initial velocity, its future motion is determined. If the size of the body is small, this motion can be described approximately by a curve. Let us now increase the mass, but not the size, of our body more and more without altering any of the other circumstances and examine how the orbit is altered by this fact.[18]

9. Field of Force; Gravitational and Electromagnetic Fields

The environment which has an influence upon the orbit is called the "field of force". It can be of two kinds. In the first case, the orbit is not at all altered by the increase of mass. An example of this case is the orbit of a projectile launched inside a vacuum. Then we call the environment "static gravitational field of force." By 'static' we mean that the field is due to masses which are all at rest (relative to the fixed stars). In the example of the projectile, for instance, the motion of the earth has to be neglected.

A typical example of the second case is the static electric or magnetic field of force. Its influence on the orbit of a charged particle becomes less and less the greater the mass becomes. We may consider, for instance, the path of an electrically charged particle moving through an electrostatic field which has a direction which is different from the initial velocity of the particle. If the charge remains constant while the mass of the particle increases, the particle will be less and less deviated by the field from its original direction. If the mass becomes "infinitely great," the particle will travel along a rectilinear path with constant speed whatever the direction of the field may be. The motion of an infinitely great mass is called an "inertial" motion, for this path is a result of mere inertia and independent of the field of force.

It is obvious that a rectilinear motion relative to a system of reference S cannot be a rectilinear motion relative to other systems S' which are accelerated or rotational in motion relative to S. To be unambiguous, we have to qualify our statement about the path of infinitely great masses. Only if we describe its motion relative to a particular system of reference S is the path of a very great mass in a field of force of the electromagnetic type a straight line and its speed constant. Such a particular system S is called an "inertial system". Our first superficial experience makes us believe that our earth is an inertial system. But more thorough investigations (of the kind of the Foucault pendulum) show us that the earth is in rotation relative to the particular system S and that rather the system of fixed stars should be taken as an inertial system, at least approximately. If we describe the path of an infinitely great mass relative to the rotating earth, it would be a kind of helix. For each system of reference it would be a characteristic curve. The inertial system is recommended only by the simplicity of this curve (straight line).

10. Newton's First Law of Motion (Law of Inertia)

These statements about the path of infinitely great masses are the factual content of what is known as the "Law of Inertia". Obviously, this law is empirical and cannot be derived from any self-evident principle. The only element in it which looks self-evident is the assertion that the straight line is the "simplest" curve. But actually it is an observation statement, too, except that it describes psychological facts, not physical ones.

The Law of Inertia has an operational meaning only if the inertial system is described by a physical operation. As a first approximation we can say that S is identical with the system of fixed stars. Since a particular physical object (the fixed stars) enters into this law, it cannot be a self-evident statement. It is even clear that the Law of Inertia, if given this operational meaning, cannot be exactly true. For the fixed stars have their "peculiar" motions and do not form a rigid frame of reference. Therefore, the Law of Inertia in this form can be only approximately true. Einstein's general theory of relativity and gravitation attempted to replace the old Law of Inertia by a law which takes into account these peculiar motions of the fixed stars. Instead of stating that a mass travels along a straight line with constant speed, we can also state that its "acceleration" is zero. ('Acceleration' means change of speed or direction.)

If we consider a finite mass, we notice by experiments that in the same environment the smaller the mass, the greater the acceleration. An electrically charged particle moving in an electric field is the more deviated from its original course the smaller its mass. (The charge belongs to the environment.) We can state this experimental fact quantitatively by stating that the product mass times acceleration remains constant if the environment does not change. Mathematically: If we denote the mass by 'm' and the acceleration by 'a' and put the product ma equal to f, the quantity f depends upon only the distances of the moving mass from the bodies of its environment.

The law $ma = f$ is known as "Newton's Second Law of Motion". We call f the "force acting upon the body m" and exerted by the bodies of the environment. If all this is true, we can derive a new operational definition of mass: The ratio of two masses is inversely proportionate to the accelerations which they get from one and the same force. This definition is only unambiguous if we obtain the same value of mass whatever force we apply. Again a physical law has to be regarded as valid in order to insure the identical result of the operations which define mass.

11. Heavy Mass and Inert Mass

The values of masses obtained from our original definition (by weighing) and our new one (by the ratio of accelerations) are identical. This is an important physical fact, too. We call "heavy mass" the result of weighing and "inert mass" the ratio of accelerations. Then we can state the law of the identity of heavy and inert mass. If we use this law, we can apply the relation $ma = f$ also to motions in gravitational fields, although it would seem that in this case the acceleration is independent of the mass. Here we have to assume that the force of gravita-

tion is proportional to the heavy mass m. The force f is equal to mg, where g is independent of m. Then $ma = f$ becomes $ma = mg$ and $a = g$. The acceleration a is independent of the mass m and is not altered if the mass becomes infinite. In this way Newton treated the gravitational field as an example of the general field of force. He advanced the hypothesis that the gravitational force is proportional to the inert mass m. This hypothesis is another way of stating the identity of heavy and inert mass. However, we must not fail to understand that the operational meaning of the terms 'inertia' and 'force' in the gravitational field is a particular one. An electromagnetic force produces an acceleration; this means a departure from the inertial motion, while in the static gravitational field the inertial motion (the motion of an infinitely great mass) itself is accelerated. The definition of force by "acceleration" and by "departure from the inertial motion" give now different results. A launched projectile is not deviated by a force if a "force" is defined as causing its departure from the motion of an infinitely great mass.

12. Newton's Second Law of Motion

Discussing Newton's Second Law of Motion ($ma = f$), the question has been raised frequently whether this law states physical facts which can be checked by experiments or whether it is only a definition of the term 'force'. Evidently, whatever may happen in the world of our experience, we can always introduce a force f which obeys the relation $ma = f$. Whenever there is an acceleration a (which is an observable quantity), we may assume the existence of a force f which cannot be observed itself but has to be calculated from the definition $ma = f$. Therefore, this "law" seems to be only a definition of the term 'force' and can be neither confirmed nor refuted by experiments. However, it is obvious that certain physical effects must be confirmed to make $ma = f$ an unambiguous operational definition of f. The product ma has to be independent of the mass m and has to depend only on the situation of the moving body in its environment. Then the law of motion says that equal masses have, under equal relations to the environment, equal accelerations. But the factual content of this law is very poor unless an operational definition of the term 'equal relations' is given which does not contain the acceleration produced.

The most famous example which decided the success of Newton's laws of motion was his theory of gravitation. The relation to the environment is determined in this case by the masses M of this environment and the distances r of the moving mass m from these masses. Every mass M contributes the term $f = Mm/r^2$ to the force. If we substitute this expression in the law $ma = f$, we obtain a law which contains only observable quantities. By using the "calculus" (Newton's equations of motion), the motion of our mass m relative to the inertial system can be calculated. This theory has succeeded in giving a very good account of the observed motions in the planetary system.

However, this success cannot be interpreted as a confirmation of the law $ma = f$ by itself but is a confirmation of this law plus the law of gravitation Mm/r^2. The meaning ascribed by Newton and the Newtonian school to the law $ma = f$ itself has been the following: Since the substitution of Mm/r^2 turned out to be such a success, it seemed probable that every motion in the universe could be described by substituting for f in $ma = f$ a formula which is in a certain way analogous to Mm/r^2. But it may contain a different power of r or even a different function of r or even the velocity of m besides the distance. The main point was that f should be expressed by a simple function of the variables which determined the relation of m to its environment.

13. Inertial Force and Centrifugal Force

Newton's Second Law of Motion, $ma = f$, where f is the gravitational or electromagnetic force, holds only if a is the acceleration with respect to an inertial system (e.g., the fixed stars). If we want to calculate the acceleration a of a mass m relative to a system which has an acceleration or a rotation relative to the fixed stars, we have to use a modification of Newton's laws. We may call the rotating or accelerated system the "vehicle" (comparing it to a railroad car). The acceleration of our mass m relative to the vehicle may be denoted by 'a_{rel}' and the acceleration of the vehicle (speaking exactly, of the point of the vehicle coinciding with the mass m) by 'a_{veh}'. If both these accelerations have the same direction as the acceleration a of our mass m relative to the fixed stars, we have obviously the equation: $a_{rel} + a_{veh} = a$. Newton's equation becomes then: $ma_{rel} = f - ma_{veh}$.

This means: The acceleration relative to the vehicle can be calculated as if it were at rest (relative to the fixed stars), except that we have to add to the electromagnetic or gravitational force f a new kind of force $f_{in} = -ma_{veh}$ which we call "inertial force". It increases proportionately to the acceleration of the vehicle but has the opposite direction. The simplest example is the shock which we feel when a railroad car is starting or stopping. a_{veh} is the acceleration of the car. If the vehicle is rotating, we can no longer use our simple relations between the accelerations.

But these relations still hold for the components of the accelerations in a certain direction. If we choose a direction perpendicular to the axis of rotation, $f_{in} = -ma_{veh}$ becomes what one calls "centrifugal force". It follows that the acceleration a_{rel}, perpendicular to the axis of rotation, can be calculated by adding to the radial component of f (electromagnetic or gravitational force) the centrifugal force. There is sometimes a confusion about whether the centrifugal force is really a "force". The answer is simple: according to the original Newtonian definition ($ma = f$), only forces of the type of gravitational or electromagnetic force can be substituted for f. The centrifugal force does not produce any acceleration with respect to the fixed stars. But if we want to describe motions relative to a rotating system, we have to add the centrifugal force f_{in} to the gravitational or electromagnetic force f.

14. The Law of Force Must Be "Simple"

If the force f is not a simple function of the distances, masses, etc., the law $ma = f$ loses its factual content. This becomes obvious if we consider what has been the most outstanding achievement of Newton's laws. We remember that Copernicus described the planetary orbits by circles, Kepler by ellipses. But if we examine the departures from the earth's elliptic orbit produced by the attraction of other plants, by Jupiter, Mars, etc., we obtain, as the effect of these perturbations, curves of an extreme complexity. Newton, however, succeeded in describing these same orbits by the simple statement that the acceleration with respect to the fixed stars is a sum of terms each of which has a simple form Mm/r^2.[19]

This means: A very complicated curve or a very complicated mathematical function can be derived from the extremely simple function Mm/r^2. If this expression of force were as complicated as the equation of the curves performed by the planets, there would be no point in replacing the geometrical description by the dynamical description. The hope of the Newtonian school has been to find for every type of motion a particular law of force of a simplicity similar to the law of gravitation—a law for the forces of cohesion, of chemical affinity, etc. This hope has been partly disappointed but partly also more than fulfilled. Since the turn of the century (1900) it has become more and more evident that the motion of the smallest particles (electrons and nuclei) cannot be described adequately by the Newtonian pattern. In the mechanics of sub-atomic particles (quantum mechanics) the term 'acceleration of a mass at a certain point in space' does not occur at all. Therefore, the concept of "force" loses its original operational meaning.

But, on the other hand, within the domain in which Newton's pattern of description can be applied at all (medium-sized bodies) not so many different laws of forces have been needed as Newton's school had expected. Besides the gravitational force, which is not even a force in the full operational meaning of this word, only electromagnetic forces have been used in twentieth-century physics—in the whole domain of phenomena which are covered by Newton's laws of motion. There are no special forces which are responsible for the phenomena of cohesion or of chemical valence. The force f in Newton's law $ma = f$ can always be calculated from the laws of electromagnetism and has therefore also an operational meaning which is independent of its definition by $ma = f$. Partly, however, these phenomena of chemical affinity and physical cohesion have to be treated by laws which are different from Newton's laws of motion. We are going to learn later that in these parts of physics "forces" are not defined by acceleration. This holds, in particular, for the phenomena of nuclear physics.

In engineering physics some laws of force are used which are "rules of thumb". In aerodynamic engineering, for instance, a law is used which says that the resistance of air which slows down the speed of an airplane is proportional to the square of its speed. But the physicists believe that this law can be derived from the laws of collisions between the molecules of the air and the molecules of the plane. The forces which are responsible for the effects of these collisions are the electric attractions and repulsions between the electric charges of which ultimately the atoms of gases and solid bodies consist.

15. Do Forces "Exist"?

These considerations will help us a little toward appreciating some attempts which have been made to apply Newton's laws of motion to living organisms by a kind of short cut. Quite a few authors have suggested introducing "vital forces" or even "spiritual forces". This would mean substituting these forces for the symbol 'f' in Newton's laws. We have seen that Newton's laws of motion have been a success in physics only because it has been possible to substitute for 'f' an expression which is simple and has a certain similarity to the law of gravitation. If one had to introduce "entelechies" or "holistic tendencies" as state variables instead of distances and masses, the whole meaning of Newton's laws would be lost. We would be faced by a completely new type of laws. To apply Newton's laws in their real scientific meaning to the motions of living organisms means to assume that the forces determining these motions are not very different from the electromagnetic forces if we remain in the domain of bodies of such a size that terms like 'acceleration' and 'velocity' have an operational meaning. A spiritual theory of the living organism perhaps does not contradict Newton's laws but would make them a definition of the term 'force' and turn them into tautologies. The opinion that such vital or spiritual forces can be introduced into Newton's laws has its source in some connotations of the term 'force'. Since this word is used in our everyday language as a term of psychology, the impression has been produced that by virtue of this term a "psychical" element has a legitimate place in physics.

As a remainder of the organismic philosophy of science which was prevailing in the Middle Ages there has been a widespread reluctance to accept the operational definition of force by which force is defined in terms of motion.

It has been argued that in a stretched piece of rubber there is a force, a tension, even if the rubber is at rest. Therefore, we are told, 'force' cannot be defined by motion and does exist as an entity independent of motion. But the statement, "Force can exist without acceleration", is a very misleading way of stating the facts. It is in flagrant contradiction to Newton's definition of force. What really happens in the stretched rubber is a state of equilibrium between several forces. The algebraic sum of the forces acting upon any particle of rubber is zero. To say that forces "exist" in this rubber would be as correct as to say that in the number zero the number five "exists" because five minus five is equal to zero. The existence of a tension or force within the rubber means in terms of operations that, by removing a part of the rubber, the equilibrium is disturbed. Acceleration of the rubber particles is produced and the force reveals itself. By saying that this force has existed before and was only balanced by other forces, we do not add anything to the operational meaning of our description of facts.

The only logical, sound way of putting the problem of the "existence of force" is again to avoid what Carnap calls the "material mode" of speaking and to stick to the "formal mode." We have to ask: If we set up a system of axioms from which Newtonian mechanics can be derived, is it necessary to introduce 'force' as an undefined term? The answer is clearly in the negative. We can introduce 'force' as a term defined by acceleration if we assume the validity of some statements about physical facts.

The insistence upon the use of the term 'force' as an "entity of its own" has its sources only in some psychological connotations of this word. We can quote the example of the Nazi philosophy in which the word 'force' is regarded as dear to the mind of the Nordic race. Every attempt to introduce a definition of 'force' by motion is branded as an act of the enemies of the Nordic race.

However, it is possible to give an operational meaning to the statement that "forces are physical realities". According to P. W. Bridgman, a term of physics describes a physical reality if several independent operational definitions of this term can be given which render one and the same numerical result.[20] If in a specific case a force cannot only be defined by acceleration but also as a function of masses, distances, etc., we have two independent operational definitions of force. We can say in this case that the term 'force' refers to a physical reality. This is obviously the case for the gravitational and electromagnetic forces which can be calculated from masses, electric charges, etc. But it is obviously not the case for spiritual and vital forces which cannot be calculated from observable data.

IV. Heat, Irreversibility, and Statistics

16. Recoverable and Irrecoverable Changes of State

If a small part of the physical world passes from a state s_0 into a state s_1, the following question arises: Is it possible that in this part of the world the original state s_0 be restored without changing anything in the rest of the world? If this restoration is possible, we call the transition from s_0 to s_1 a 'recoverable transition'. If not, we speak of an 'irrecoverable (irreversible) change of state'. To understand the role of irrecoverable phenomena and in particular the phenomena of heat, we are going to consider a simple example.[21]

The part of the world which we consider may be a perfect gas which is isolated from any supply or loss of heat. But we can change the volume of this gas by moving a piston closing the container of the gas. The initial state s_0 is a certain volume V_0 and a certain temperature T_0 of our gas, e.g., a volume of 2 liters and a temperature of 0° centigrade. The second state s_1 may be the result of an increasing pressure upon the piston. The final volume may be V_1 (e.g., 1 liter). We perform the compression by putting only enough weight on the piston to make the pressure exerted just a little greater than the expansion pressure of the gas. Then the transition from the initial volume V_0 to the final volume V_1 will occur infinitely slowly. Since the gas is, during the whole transition, almost in equilibrium, we speak of a quasistatic change of state. This means that the process is almost static. As no heat is lost, the temperature increases by the compression. When the gas has reached the final volume V_1, the temperature has increased to a value T_1, which is greater than T_0. According to an easy calculation, this temperature can be calculated by the formula: $T_1 V_1^{k-1} = T_0 V_0^{k-1}$ (k has the value 1.4 for a diatomic gas like oxygen).

Outside our gas the effect of the process has been the lowering of the weight which has been pressing upon the piston. Hence, the effect on the whole world has been lowering of a weight, increase in a temperature, and decrease in the volume of our gas. Is this change of state from s_0 to s_1 recoverable? Can the state s_0 be restored? Certainly, yes. We have only to reduce slowly the pressure upon the piston. Then the gas will expand again infinitely slowly because the pressure on the piston is just a little smaller than the internal pressure of the gas. When the original volume V_0 is restored by expansion, the weight on the piston is in its original position and the initial low temperature T_0 is also restored. This means that the change of state from s_0 to s_1 is really recoverable.

Let us now modify our change of state. We start again from s_0 but produce a change of state in a new way. The gas may remain isolated against

heat supply from outside. Now we increase its temperature. But we do it at constant volume V_0. We can, e.g., like in the old Joule experiment, have a paddle wheel rotating in the gas. The rotation can be produced by the fall of a weight which is connected with the wheel by a cord over a pulley. The gas will be heated by friction. The state s_0 will be changed into a state s_2 which has the same volume V_0 but a temperature T_2 which is greater than T_0. Is the change of state from s_0 to s_2 now also recoverable? Can a change of state occur which restores s_0 if no heat is supplied from outside? This change would mean that the weight which produced the friction by falling has been lifted again, while the temperature of the gas which had increased by friction has dropped again. The final result of the change of state from s_2 to s_0 would be the decrease in temperature of the gas and the lifting of a weight outside the gas, while nothing else has changed in the world.

If we generalize the results of all our experiences on the conversion of heat into work, we notice that a weight can never be lifted just by using the heat supplied by the drop in the temperature of one single reservoir. This generalization has become a basic hypothesis of physics. It is usually called the "Second Principle of thermodynamics" (the First Principle states the conservation of energy). From this Second Principle it follows that the change of state from s_0 to s_2 is not recoverable. This means: The increase of temperature by friction at constant volume is not recoverable. It is obvious that by this change of state from s_0 to s_2 (friction) the function TV^{k-1} increases as V remains constant and T increases.

17. The Second Principle of Thermodynamics and Entropy

The function $F = TV^{k-1}$ increases by the process of friction and would decrease if a weight were lifted and nothing else happened than the decrease in the temperature of one body (e.g., our gas) at constant volume. The function $F = TV^{k-1}$ may, provisionally, be called "entropy". This entropy remains obviously unaltered as long as the process is quasistatic and no heat is supplied. The entropy increases if a change of state takes place which is irrecoverable. The entropy would decrease during a change of state which is forbidden by the Second Principle of thermodynamics. This principle can therefore be formulated for a perfect gas as follows: There is a function F of the state variables V, T with this property: if no heat is supplied, F can only increase. It remains constant during a quasistatic process. But this is only a limit which can never be reached by an actual experiment.

If we pass from a perfect gas to a general system, we can formulate the Second Principle in an analogous way. The whole system consists of several bodies with the volumes V_1, V_2, etc., and the temperatures T_1, T_2, etc. Then there is again a function $F(T_1, T_2, \ldots, V_1, V_2, \ldots)$ of the state variables which behaves in a simple way during changes of state which are not accompanied by an exchange of heat between these bodies and the outside world. The function F remains constant during quasistatic processes. These changes of state are recoverable with the approximation with which a quasistatic process can be performed at all. All changes of state which entail an increase of F are irrecoverable, while changes of state with decreasing F are impossible.

As a matter of fact, every function of F which is increasing with increasing F has the same properties as the "entropy". It turns out to be convenient not to introduce $F = TV^{k-1}$ but $S(V,T) = \log F$ as the "entropy" of a perfect gas. For if we consider two gases with the common temperature T and the volumes V_1 and V_2, we can show easily that during a quasistatic process

$$F\,(V_1\,,T) + F\,(V_2\,,T) = TV_1^{k-1} + TV_2^{k-1}$$

does not remain constant while

$$\log TV_1^{k-1} + \log TV_2^{k-1}$$

does. If we introduce the function $S = \log F$,

$$S\,(V,T) = \log TV^{k-1} = \log T$$
$$+ (k-1)\log V\,,$$

and call it definitely "entropy", we can assert a simple law. If S_1 and S_2 are the entropies of two gases which are in temperature equilibrium, the entropy of the whole system is $S_1 + S_2$. This means that $S_1 + S_2$ remains constant during quasistatic processes of the whole system.

The most popular example of increasing entropy is the passage of a system in which there are great differences in temperature into a system in which these differences are smaller. In the case that no heat is supplied from outside, one can show by an easy calculation that the entropy increases by the disappearance of these temperature differences. Therefore, a process of leveling temperature intensity in an isolated system is irrecoverable. Only in the borderline case of a quasistatic process does the entropy of the isolated system remain constant. The system performs recoverable changes of state. Therefore, the quasistatic processes are also called 'reversible processes'.

18. Cosmological Implications

All these facts and generalizations have been used to predict a very gloomy future for our world. The Second Principle of thermodynamics has been formulated as follows: The entropy of the universe tends toward a maximum, and this maximum is attained if the whole world is of constant temperature. Since the world is an isolated system, every change of state is connected with an increase of entropy. When the world has reached the state of greatest entropy, no change is longer possible; the universe will "die".

In a similar way it has been argued that only a finite number of years have passed since the creation of the world. If the actual laws of nature had been valid through an infinite number of years, the maximum of entropy would have already been attained. Since this is obviously not the case and the world is still "alive", the laws of nature and our world itself cannot have existed for an infinite number of years. It must have been created a finite number of years ago.

Since all processes which are covered by Newton's laws of motion can be reversed, this tendency of the world toward the ceasing of all changes has been interpreted as a feature of nature which cannot be explained by mechanics. A kind of striving toward an end has been envisaged by many philosophers. It has been cheered as a spiritual factor. This running toward the death of the world has been interpreted as being in contrast to the evolution of organisms which shows a tendency toward greater and greater differentiation. Two conflicting tendencies must therefore be recognized in the universe both of which cannot be explained by mechanical laws. Sometimes this conflict has been interpreted as a scientific background for the eternal struggle between God and the devil.[22]

As a matter of fact, the term 'entropy of the universe' is, as P. W. Bridgman says correctly, a pure "paper and pencil affair". We can, of course, say that the entropy of the universe is the sum of the entropies of all parts of the universe. But speaking exactly we do not know whether the expression 'entropy of the Universe' has any operational meaning. For the sum of the entropies of the parts of a system can be regarded only under very restricted conditions as the entropy of the whole system. We do not even know whether the sum is convergent from the purely mathematical viewpoint. Therefore, all these conclusions concerning the creation and death of the world are results of a loose way of thinking.[23]

19. The Kinetic Theory of Heat

We shall understand the whole problem of recoverability much better if we attack the question from a different angle. We ask whether "irrecoverability" is compatible with the hypothesis that heat is a movement of small particles. Then we shall learn that irrecoverability is the result of a rather shortsighted view of the universe, while a more penetrating view would reveal to us a universal recoverability. We must never forget that the results of thermodynamics have operational meaning only under conditions which can be described by the concepts of thermodynamics: temperature, mechanical work, entropy, etc. But there are conditions which cannot be described by what P. W. Bridgman calls the "universe of operations" of thermodynamics. This fact is obvious from the simple consideration that in no statement of thermodynamics is the expression 'velocity' or 'time' used. Therefore, we are given not the slightest estimation of how long it may take to establish a certain state of the world. If a theory does not contain any statement about time, it has no operational meaning to say that the

theory predicts a certain future of the world. For the time may just as well be "never."

There are certainly phenomena of heat which are not covered by the laws of thermodynamics. In this category belong all phenomena of fluctuation like the Brownian movement, the spontaneous change of density in the atmosphere which accounts for the blue sky, etc. For this reason we are in need of another method to deal with the phenomena of heat. This method is provided by the "kinetic" or "statistical" or "molecular" theory of heat. Its basic hypothesis asserts that, besides the observable motions of bodies, there is an irregular zigzag movement of its microscopic and submicroscopic particles which on the average does not contribute anything to the observable velocity of medium-sized bodies. Moreover, according to this hypothesis, the average kinetic energy of this molecular motion is proportional to the absolute temperature of the body.

This hypothesis which accounts for a great many facts connected with the conversion of heat into mechanical work and vice versa faces a great difficulty if we apply it to the existence of irrecoverable changes in the universe.

20. Mechanics Cannot Account for Irrecoverable Changes of State

We consider a simple case. There may be a gas inclosed in a container. The gas molecules are performing rectilinear motions and are only deflected if they hit each other or the walls of the container. We start from a state s_0 in which all molecules are assembled in a small corner of the container while most of its volume is empty. This means in terms of thermodynamics that the density of the gas in this corner is great while the density in the rest of the volume is zero. In the second state, s_1, the molecules may fill up the whole container with equal density. According to the rules for calculating the entropy, this physical quantity has in the state s_1 a much greater value than in the state s_0 of our gas. As in an isolated system the entropy can only increase, the change of state from s_0 to s_1 is, according to the laws of thermodynamics, irrecoverable.

According to these laws, it can never happen that a gas which fills up a volume with constant density can undergo changes of density by which eventually the whole mass is assembled in one corner. However, according to the kinetic theory, we cannot understand why such a thing should be impossible. Let us assume that the state s_1 (equal density) has just been reached. Then we reverse the direction of the velocities of all particles. According to Newton's laws of motion, they must traverse now the same orbits as before but in an opposite direction, until they reach finally again the state s_0 and assemble in one corner. Therefore, the kinetic theory seems to be incompatible with the existence of irrecoverable changes of state. This means that the kinetic theory would be incompatible with the Second Principle of thermodynamics. For from mechanics it

seems to follow that a homogeneous gas can spontaneously move into a small corner.

This argument is conclusive. Newton's mechanics as the general Law of the Universe and the existence of irrecoverable changes are incompatible. The kinetic theory of heat cannot be based on mere mechanics. It needs, in addition to Newton's laws of motion, still a different type of hypothesis. The kinetic theory makes use of *statistical hypotheses*. L. Boltzmann was the first to point out how, by combining mechanical and statistical hypotheses, it can be predicted that the initial state of leveling processes like heat conduction and diffusion is "irrecoverable", if we take this word in its practical sense. However, it can also be derived from statistical hypotheses that this irrecoverability is only a superficial aspect. If we observe the phenomena more in detail and over long periods of time, we can bet that the initial state will once reappear.

21. Statistics and Irrecoverability

The statistical hypotheses, in contrast to the mechanical hypotheses, do not make an assumption about how a specific particle behaves at a specific time in a given field of force. Statistical assumptions say, in a certain sense, more and, in a certain sense, less. We explain their meaning by a simple example. We may divide the whole volume of our container of gas into a great many small cells. Our hypothesis says: If we pursue the path of a particle over a long time (years or centuries), the time during which it dwells in a specific cell is a specific fraction of the whole time considered. The volume of the cell is in the same ratio to the whole volume of the container as the dwelling time in the cell is to the whole time considered. This means that this ratio becomes more and more independent of the time of observation. This ratio may be denoted by 'p'. If there are, e.g., 100,000 cells, p is a very small fraction (e.g., $p = 1/100,000$) and may be called the "dwelling time" in a specific cell. If we consider a second gas molecule moving independently of the first one, it spends again the $1/100,000$ part of any given time stretch in our specific cell. If T is the whole time of observation and T' is the dwelling time of the first particle in our cell and T'' the time during which both the first and the second particle are dwelling in our cell, we have

$$T' = pT \qquad T'' = pT' = p^2T \ .$$

If we want to know the time during which N particles would all dwell in one specific cell, we would find that it is equal to $p^N T$. If we assume, e.g., that N is equal to one million (10^6), the dwelling time of all N particles in one specific cell would be

$$T \cdot p^{1,000,000} = T \left(\tfrac{1}{100,000}\right)^{1,000,000}$$
$$= T \ \frac{1}{10^{5 \times 10^6}} = T \ \frac{1}{10^{5,000,000}} \ .$$

The denominator is a number in which the digit 'one' is followed by five million zeros. Therefore, if we make ob-

servations during a billion years ($T =$ a billion) the average "dwelling time" of our particles in one particular corner will be much less than the billionth of the billionth part of a second. Moreover, if we are the lucky observers of such an event, we can bet that it will soon disappear and not reappear for a billion generations to come.

This is the statistical aspect of irrecoverability: There are rare states of a system and they disappear very soon. Hence, we can bet in what direction the system will develop. But if we start from a very frequent state (e.g., from the state where the particles are equally distributed over the whole container), we can bet that it will hardly happen in a billion generations that the rare state will be restored in which all particles assemble in a particular corner.

The sequence "rare state–frequent state" happens as often as "frequent state–rate state". The illusion of irrecoverability in the realm of our observations results from the fact that we start every experiment from a "rare" state of our system and our time T is always short. Then we can bet that the future will bring states which are not so rare. As a result, essentially there is no asymmetry of sequence but a distinction between rare and frequent states of a system. Starting from a rare state, s_0, the next state will usually be a frequent one, while, starting from a frequent state, the next one will usually be again a frequent one.

If we pass from the special case of a gas to a general case, we may say that in every system we have to set up a statistical hypothesis which allows us to compare the different states of a system with respect to their relative dwelling times.

22. Statistical Hypotheses Are Assertions about Physical Facts

The kinetic theory of heat is based upon statistical hypotheses which make predictions about motions of particles as the mechanical hypotheses do. They could, therefore, contradict Newton's laws of motion. In setting up these statistical hypotheses, their compatibility with Newton's mechanics has to be assumed as a special hypothesis. Every statistical hypothesis assumes a specific regularity about the dwelling time of particles in a certain domain. The essential in "Boltzmann's statistics" is the assumption that the dwelling time of a particle in any specific domain is independent of how many other particles are present in this domain. If we drop this hypothesis, we alter substantially the conclusions drawn from the kinetic theory. In the modern wave mechanics, which has replaced Newton's mechanics, Boltzmann statistics have been replaced by other statistical hypotheses. For each particle a particular statistical hypothesis has been set up. For some particles (e.g., photons) the dwelling time of the particle is increased by the presence of other particles (Bose statistics). In other cases (e.g., electrons) the dwelling time is unfavorably influenced by the presence of other particles (Fermi statistics).

We must not be too puzzled by the fact that irrecoverability only comes into the picture if we start from a rare or improbable state of the system. But we may ask: How does it happen that we are living in a period and on a place of the universe where the entropy is increasing? We must always remember that the statistical approach to the happenings in the universe starts from the assumption that these happenings are in agreement with our basic statistical hypotheses. This means that the universe has already traversed a great many cycles. All the rare states have appeared and disappeared and will reappear again. This universe is, in the sense of ancient philosophy, an Epicurean universe. The origin of the sun, the earth, the elements, and even of our own human race is due to the law of chance. The appearance of men in the universe is a very improbable event from this viewpoint. In the Aristotelian universe, on the other hand, everything has developed according to plan toward a certain end of higher perfection. However, if we start from the assumption that our universe is an Epicurean one, our whole existence is due to chance, and we, the human race, are a very improbable event. Therefore, it is obvious that the universe will mostly go back to a state which appears more frequently according to the fundamental statistical hypothesis. But all the rare states have their proper chance to reappear.

These considerations are pertinent for the attitude of philosophers toward the principles of thermodynamics. A great many philosophic interpretations of physics have made use of the principle of increasing entropy to bolster up an anti-mechanistic teleological view of the universe which invokes a tendency, a direction toward a certain end, instead of a causal chain of events. Unfortunately, the trend of events, if we take thermodynamics for granted, tends toward destruction of the universe. If we, on the other hand, replace pure thermodynamics by the kinetic theory of heat (laws of motion + statistical hypotheses) we drop implicitly the Aristotelian theory of the universe and accept the Epicurean view that every tendency toward an end is an illusion and the real actor in the evolution of the world is a play of chances and the survival of the fittest. There is no real irrecoverability.

V. The Theory of Relativity

23. Einstein's Two Basic Principles

Newton's mechanics was applied to a hypothetical medium, the "ether", which was supposed to fill up the whole world-space. The wave motion in this ether was interpreted as being responsible for what appears to our sense observations as phenonema of light. The "ether theory of light" treats these waves according to Newton's laws of motion. The first result of this theory is that in empty space the waves of light are propagated with a speed $c = 3 \times 10^{10}$ cm/sec with respect to the ether. The speed of the source of light has no influence upon this speed of propagation. This result can also be formulated without speaking of the ether. One proceeds in a way which is similar to the replacing of Newton's absolute space by the "inertial system" (sec. 9). We say that in empty space the waves of light are propagated with the speed c relative to a certain "fundamental system S".

Let us now examine phenomena due to the simultaneous motion of material bodies and propagation of light waves. Then Newtonian mechanics and optics lead to results which are not in agreement with the experimental facts and, moreover, are by themselves rather awkward. If we stick to the belief that Newton's laws of motion are the only legitimate basis of physics and that light is in particular a wave motion in a medium which follows these laws, we have to resort to complicated additional hypotheses.

In 1905 Einstein dropped both assumptions: Newtonian mechanics and the ether theory of light. His new theory of light and motion retained Newton's laws only for a limited class of motions which obviously follow these laws. These are the motions with "small velocities." This means velocities which are small in comparison to the velocity of light. Troubles arise only if the speed of material bodies approaches the speed of light. Further, Einstein retained two results of the ether theory of light which are made plausible by experiments. But he regarded these results as straight generalizations of experimental facts without troubling about whether they could be derived from the ether theory of light or from Newton's laws of motion.

Einstein's two hypotheses (or principles) are: first, the speed of light relative to any room is independent of the speed of the source of light relative to this room (principle of the constancy of light velocity); second, the propagation of light relative to the walls of a moving room is determined by the initial conditions relative to this room provided that the motion of this room is a rectilinear and uniform motion relative to the fundamental system S (principle of relativity).

This principle can also be formulated as a statement about the impossibility of performing a certain set of operations. If our room has the speed v relative to S, it says that there is no physical experiment by which we can measure this quantity v, i.e., the expression 'speed of a room relative to the fundamental system' has no operational meaning. If we substitute into Einstein's two principles the operational definition of 'speed of light relative to a system', 'speed of the light source', etc., they become physical hypotheses about the interactions between the motion of bodies and the propagation of light. They are to be confirmed as any physical hypotheses by the agreement of their results with direct observations. These results are in conspicuous contradiction to the results derived from Newton's laws of motion in connection with the traditional ether theory of light. For this traditional theory allows for a particular operation by which this speed v can be determined (Michelson's experiment).

24. Newton's Laws of Motion Not Universally Valid

Actually, none of the observable phenomena derived from Einstein's two principles has been in disagreement with the results of observation. In this sense all statements derived from these principles are "true". Then it follows easily that some immediate results of Newton's laws of motion cannot be true. According to these laws, the increment of velocity produced by a given force is independent of the actual (initial) velocity. Therefore, by the continuous action of a constant force, any increment of velocity could be produced. We could, e.g., obtain in this manner a velocity v of a body greater than the velocity c of light in a vacuum. But we can show easily that this result is in flagrant contradiction to Einstein's principle of relativity. Assuming the universal validity of Newton's laws, we could consider a room which is moving with the speed of light relative to S. By a light source at rest in this room light is emitted through the vacuum toward one of the walls. This light will never reach this wall which itself moves with the speed of light. Therefore, no reflection can occur in this direction. By the absence of reflection we could make sure that our room has the speed c relative to S. But the determination of this speed is impossible, according to the principle of relativity.

Therefore the assumption that Einstein's principle is valid implies the negation of Newton's laws. We must assume that in contrast to Newton's mechanics the increment of velocity produced by a force is dependent upon the actual velocity. We can calculate exactly this dependence from Einstein's principles and obtain the result that this increment becomes the smaller the greater is the actual velocity, and tends toward zero if the actual velocity approaches the speed of light (the force being constant). This modification of Newton's laws can be checked by experiments. One can observe the increment of velocities produced by electric and magnetic forces if one examines fast-moving electrons, as cathode rays or the rays emitted by radioactive substances. These rays travel with a speed which is comparable to the speed of light. According to Einstein's results, the dependence of the increment of velocity upon these great initial velocities is great enough to be observable.

25. The Operational Definition of 'Mass' Has To Be Modified

This dependence has been confirmed by experiments. Hence some operations which are to render an identical result according to Newton's laws no longer do so. Let us consider the operational meaning of 'the mass of a particle'. If we assume that the field of force is known (e.g., given by Coulomb's law), we can get the mass by measuring the acceleration a (increment of velocity per unit of time). The mass m is defined by $m = f/a$. According to Newton's mechanics, the result is independent of whether the initial velocity was small or great. But if Einstein's principles are right, this operational definition becomes ambiguous. The acceleration a (and, therefore, m) depends actually upon from what initial velocity we start the experiment. In order to obtain an unambiguous result, we have to specify the operations involved, in particular the initial velocity v. If we require that the initial velocity be zero relative to S, the acceleration becomes unambiguously determined. We must therefore use a modified operational definition of 'mass'. We can either make the specification that the initial velocity relative to S is zero; then we define a concept which is called "rest mass" m_0. Or we can include the initial velocity v in the description of the operation. Then acceleration and mass themselves become dependent upon v. We obtain a physical quantity which is no longer a constant but a function of v. This quantity is called "mass" m in the new mechanics. By using this definition, we can formulate the laws of motion in the simple form: mass times acceleration equals force ($ma = f$). But the mass m is now a function of v.

26. The Operational Definition of 'Time Distance' Becomes Ambiguous

If we assume that all observable results of Einstein's two principles are true, we have to assume that all results which can be logically derived from them are also true, even if they are not observable directly. Among these conclusions are several which have been exciting to a great many people. They have seemed to contradict common sense, in general, and the glorified common sense called "philosophical insight", in particular.

We can conclude, e.g., from Einstein's two principles that a clock which travels with the speed v relative to S loses time compared with the clocks at rest in S. If we recall the operational definition of a clock, we notice again that some operations which rendered, according to Newton's laws, identical results no longer do so if Einstein's principles are assumed to be true. The operations by which the time distance between two events was defined did not mention the speed of the clock relative to any system of reference. For, according to Newton's physics, this speed is without influence upon the march of the clock. If Einstein's principles are true, this operational definition of the time distance between two events becomes

ambiguous. We must specify the speed of the clocks used in this measurement. In order to obtain an unambiguous result of our defining operation, we must no longer say that "between the events A and B there is a time distance of 10 seconds" but that "there is a time distance of 10 seconds if we use clocks which are at rest in a particular system S'". We can express this statement a little more briefly by saying that "this time distance is 10 seconds relative to the system S'". The velocity of S' relative to S must be specifically given. We use again a "relativized language" in order to make the description and the operations unambiguous.

By starting from Einstein's principles, one can derive new laws of motion which differ for high speed from Newton's laws. We can in the same way derive laws for the propagation of light through material bodies which could be derived from Newton's mechanics and traditional wave optics only by complicated additional hypotheses.

The system of all statements which can be derived from Einstein's principles is called "the Theory of Relativity". This theory is so called because its characteristic and basic hypothesis is a principle of relativity. Before Einstein the validity of Newton's laws of motion and of the theorem of relativity for optical phenomena seemed to be incompatible. Einstein's master-idea was to drop Newton's laws of motion as well as the ether theory of light and to generalize

Newton's theorem of relativity into a general hypothesis which should be valid in the whole domain of the motion of material bodies and of light propagation.

27. The Relativity of Time

By people who are interested in the philosophical aspects of relativity theory the question has been raised: Has science really proved that time is relative and not absolute? The answer can be given quite directly.

The statement that "time is relative" can have two different meanings. First, it may mean that the march of a clock is altered by a rectilinear motion with constant speed v. This assertion is a statement about facts which can be checked directly or indirectly by experiments. It is confirmed indirectly by every experiment which confirms an observable result drawn from Einstein's principles. It is checked more directly by experiments which examine how the frequency of the light emitted by a sodium atom is altered by the motion of this atom (atomic clock, Ives's experiment).

But this physical meaning of the relativity of time is not the one which has puzzled the philosophers and the great public. 'Relativity of time' has a second meaning—a logical or philosophical one. From the theoretical and experimental results of Einstein's principles it becomes evident that the statement that "there is a time distance of 10 seconds between two given events" has no operational meaning.

We must always add explicitly a particular system of reference. Therefore, Einstein suggested dropping entirely sentences of the form "the time distance is 10 seconds" from the language of physics and of admitting only sentences of the form "the time distance is 10 seconds relative to the system of reference S". This suggestion is not a statement about physical facts which can be confirmed or refuted by experiments. It is a suggestion for using a language which is recommended as being well adjusted to our experience about physical facts.

Nobody can be persuaded to accept this recommendation if he does not like to. The adjustment of our language to the physical facts may not be the primary motive of our rules of language. Perhaps we want to adjust our language to tasks which are beyond our desire for a clear description of facts. If we wish, we can, of course, use the expression 'real time distance between two events' without referring to a particular system of reference. But we must know how to give to such a statement an unambiguous operational meaning. Either we concede that we do not know which of the arbitrary systems of reference is the "real" one and leave the discovery of this system to a superior spirit, or we single out some convenient system (e.g., the system of fixed stars) and give to the time distance relative to this particular system the honorary title "real or absolute time distance" while we call the time distance relative to other systems "relative or apparent time distance". Nobody can prove that this way of speaking is "false". If we believe, e.g., that our moral values should be described by using the term 'absolute', one could find it convenient to use in the description of the physical world as far as possible the same terminology as in describing the goals of human behavior. Then it would be advisable for believers in "absolute values" to use also in physics the expression 'absolute time distance.'

28. The Speed of Light Has the Same Value in All Systems of Reference

Among the results of Einstein's principles which cannot be checked directly by experiments, the one which has been regarded frequently as its "central absurdity" is perhaps the following: Any light ray has, relative to S, the same speed as relative to a system S' which moves with an arbitrary speed v uniformly relative to S. Some philosophers have taxed this statement as being absurd or even self-contradictory. They argue: The speed of light relative to S is c. If S' moves with the speed v (which is smaller than c) in the direction of the light ray, it advances more slowly than the light. Therefore, the speed of light relative to S' is only $c-v$, which is obviously smaller than c. To assert that the speed of our light ray relative to S' is also c means to say that $c-v$ is equal to c while v is different from zero. This is obviously absurd. However, through

all this argument the velocities are measured by instruments at rest in S. Only if this operational definition of speed is used, can it be proved that the speed of light relative to S' is $c-v$. But in Einstein's statement 'velocity relative to S'' means the velocity measured by instruments which are at rest in S'. Then it cannot be proved that this speed of light relative to S' is $c-v$ but rather that it is c. The confusion comes from the failure to distinguish between mathematical symbols 'v' and 'c' and their operational meaning. Without the addition of the operational definitions, physical statements do not say anything about physical facts and cannot be checked by experiments.

If we have this in mind, we can understand that the statement that "no material body can move with a speed which is equal to or greater than the speed of light" is not a statement about absolute motion. It means, of course, that no material body can move with the speed c relative to S, but it means just as well that no material body can move with the speed c relative to S' (which moves in turn with the speed v relative to S). The real operational meaning of our statement about the speed of light as the upper limit of all speeds of masses is the following: If we have a set of clocks and yardsticks adjusted to a system S, no material system S' can move with the speed c relative to S if we measure c by the instruments in S.

29. The Traveling Twins

Another startling result is the story about the traveling twins. We learned that a clock which is moving with the speed v relative to S loses time compared with the clocks which are at rest in S. A human being is a kind of clock. The chief functions of the human organism, in particular the heartbeats, are achieved by periodical mechanisms. According to Einstein's principles, every periodical mechanism behaves like a clock. If we have twins, one at rest in S and one moving with the speed v relative to S, the traveling one will experience a smaller number of heart-beats. This can be seen particularly clearly if the traveling brother traverses a large circle which eventually leads him back to his stationary brother. Then the traveler must have remained younger than his brother at home. This conclusion is a necessary consequence of Einstein's principles. This result can, of course, neither be confirmed nor refuted by direct experiments. We never have to do with organisms traveling at a speed which is near to the speed of light without any violent disturbance produced by an acceleration. The possibility of this phenomenon can perhaps be understood if we consider the recent physiological experiment concerning the conservation of organisms at low temperatures. For a bath at a temperature well below the freezing-point brings about a slowing-down of the periodical processes inside the organism. It follows certainly from Einstein's two principles that the twin brother who has been at rest in the fundamental system S becomes older than his brother who performed a circular motion relative to this system.[24]

30. Conversion of Mass into Energy and Vice Versa

According to Newtonian physics, the sum of masses cannot be changed by any interaction of material bodies. According to the theory of relativity, the masses are dependent upon speed. However, the question arises whether the sum of the rest masses can be altered by interaction. From Einstein's principles it can be derived that the sum of the rest masses remains unaltered only if the sum of the kinetic energies is not altered by the interaction considered.

But if we consider phenomena which are connected with the conversion of kinetic energy into other forms of energy, e.g., into heat, the matter is different. As the simplest case we consider two bodies of equal rest masses which move with equal speeds relative to S in opposite directions. A collision takes place. If the bodies are perfectly rigid (unelastic), both bodies will come to rest in S and their whole kinetic energy will disappear and be converted into heat energy. It can be strictly derived from relativity theory that by this collision the sum of the rest masses of the colliding bodies is changed. Before the collision the rest masses of each body were m_0; the rest mass of the whole system, $2m_0$. After the collision we have one larger body consisting of both small ones. Its rest mass will be greater than $2m_0$. From the theory of relativity it can be strictly derived that the increment of rest mass will be equal to the loss of kinetic energy divided by c^2 (the square of the speed of light). Since the loss of kinetic energy is equal to the produced heat energy H, we can also say that the rest mass after the collision is $2m_0 + H/c^2$. This is a special case of a more general theorem which follows from the theory of relativity: If by an interaction of material bodies kinetic energy E disappears, the sum of the participating rest masses increases by E/c^2. If kinetic energy E is produced by the interaction, the sum of the rest masses would decrease by E/c^2.

We can verbalize this fact by saying that rest mass is partly "converted" into kinetic energy. If the rest mass decreases by Δm_0, the kinetic energy $E = c^2 \Delta m_0$ is produced. The same thing is true if we replace production of kinetic energy E by production of radiant energy E. Continuing this line of argument, one can envisage the possibility that the whole rest mass m of a body could be converted into energy. Then the energy $E = m_0 c^2$ would be produced and the whole rest mass of the body would disappear. There are phenomena in nuclear physics which seem to lend themselves easily to an interpretation of a conversion of matter into energy (such as in the atomic bomb).

A well-known phenomenon of this kind is the so-called "mass defect" of atomic nuclei. The nucleus of the helium atom, e.g., consists of four particles, each of which has the rest mass m_0 of the hydrogen nucleus (proton). However, the mass of the helium nucleus is smaller than four times the rest mass of the hydrogen nucleus. If we assume that the particles in the nucleus are kept together by attracting forces, the particles which are packed in the helium nucleus have a smaller potential energy than they had when separated from one another. The formation of the helium nucleus is connected with a loss of potential energy and therefore with a production of kinetic or radiant energy. If the rest mass of the hydrogen nucleus is m_0 and the rest mass of the helium nucleus $4m_0 - D$, the drop of potential energy accompanying the formation of one helium nucleus is Dc^2. This energy can be regarded as the "binding energy" which keeps the helium nucleus together. There are a great many cases of disintegration of an atomic nucleus where the binding energy can be measured directly. We find a good agreement with the energy calculated from the mass defect.

31. Creation and Annihilation of Mass

Another phenomenon of this type is the creation or annihilation of an electron-positron pair. As we shall learn in Part VIII, the "electron" has a much smaller rest mass than the proton and a negative electric charge.

The "positron" has the same mass as the electron, but the charge is positive. The magnitude of the charges of electron and positron is equal. If they combine, the charges disappear of course. Moreover, there is sufficient evidence to assume that the rest masses disappear also and are converted into radiant energy. This assumption is confirmed particularly by the measurement of the wave-lengths of the radiation which accompanies the disappearance of an electron-positron pair. For (according to sec. 39) radiation consists of elementary portions of radiant energy called photons. The wave-lengths λ of such a photon can be easily calculated from the energy E by the formula $E = hc/\lambda$. Roughly speaking, the smaller the energy, the greater the wave-lengths. If we calculate the loss of energy E produced by the disappearance of the rest mass of an electron-positron pair ($E = 2m_0c^2$), we obtain the wave-lengths of this radiation from $2m_0c^2 = hc/\lambda$. It is a very hard gamma radiation and is really found if one observes the phenomena connected with the disappearance of positrons.

We have learned now how the conversion of mass into energy or the annihilation of mass can be described very distinctly and clearly by describing performable physical operations. The annihilation of mass has been interpreted occasionally as a refutation of materialism and support of spiritualism. This can be done only if one uses this language without having in mind the operational mean-

ing of the words and sentences. The statement, "Matter can be annihilated and converted into energy", sounds strange to the man of average school training and smacks at least of spiritualism. For he understands tacitly the words 'matter' and 'energy' the way they are used in everyday language. In this language by 'matter' is meant something like rock or ocean. By 'energy' is meant something like a soul or a spirit. The dualistic view of the contrast between body and mind is deeply entrenched in our everyday language. In this language, 'matter' has the operational meaning of some hard and impenetrable stuff, while 'energy' seems to be defined by operations which belong in the field of psychology rather than in the field of physics.

The physicist, in particular the twentieth-century physicist, means by 'matter' a system consisting of a great number of particles like electrons, protons, positrons, etc. The operational definitions of these particles are very different from the operations by which we used to test the presence of some hard or impenetrable stuff like the "matter" of our everyday language. If we take into account the meaning of the words 'electron' or 'proton', which are used in the wave theory of matter, we are very far from operations testing hardness or impenetrability. On the other hand, radiant energy is regarded by the physicist as consisting of photons. The presence of photons is tested by methods which are essentially of the same kind as the methods by which the presence of electrons is tested. Therefore, if we introduce the operational meaning of the terms 'matter' and 'energy', the annihilation of matter and its con-version into energy is not at all obscure and has nothing to do with the experiments of the spiritualists which demonstrate "dematerialization".

VI. Light

32. The Crucial Experiments on Behalf of the Wave Theory

A century ago there were two theories of light both of which were based upon Newtonian mechanics. According to the "corpuscular" hypothesis, light consists of small corpuscles which obey Newton's laws of motion. They perform rectilinear motions with constant speed. If passing from vacuum or air into water, they are attracted by the particles of water and deviated from their original path. The deviating forces are similar to Newton's force of gravitation. From this hypothesis it can be derived that between the angle of incidence a and the angle of refraction β there is the relation $\sin a / \sin \beta = c'/c$, where c is the speed of light in vacuum (or in air) and c' is the speed of light in water. Since by the attraction the light corpuscles are accelerated, it follows that c' is greater than c; the speed of light in water is greater than in air ($c' > c$). But Foucault performed in 1850 an experiment by which he proved that the speed of light in water is smaller than in air in contrast to the claim of the corpuscular theory. Therefore, the second of the conflicting hypotheses about light propagation seemed to be confirmed: the "wave theory" of light. According to this theory, light consists of waves propagated through an elastic medium (ether) which follows Newton's laws of motion. From this hypothesis Huyghens could derive that the speed of light in water is smaller than in air ($c' < c$). This result seemed to be confirmed by the Foucault experiment.

The Foucault experiment has been quoted over and over again as the outstanding example of a crucial experiment. It was supposed to have sealed the death sentence of the corpuscular theory. But if we take a strictly logical viewpoint, we can say only that the experiment decided against a corpuscular theory of light which assumed that the corpuscles follow Newton's laws of motion. Therefore, we can speak only of a decision against any corpuscular theory of light if we mean by 'corpuscle' a small mass in the Newtonian sense of this word. It is obviously not excluded by the Foucault experiment that light consists of corpuscles in a wider sense which move according to laws which differ from Newton's laws. We have to understand that an experiment can only be crucial if only one clear-cut alternative exists. An experiment can never decide definitely between wave theory and corpuscular theory but at most between Newtonian waves and Newtonian corpuscles. By Foucault's experiment Newtonian corpuscles are outruled, but this does not mean a decision in favor of Newtonian waves, unless we make the assumption that only this alternative exists. Perhaps Newtonian mechanics has to be dropped outright.

33. Conversion of Light into Kinetic Energy; the Photon

Half a century after Foucault's crucial experiment it had become more and more clear that the wave theory of light in its original form did not suffice to cover the whole realm of newly discovered facts. The new facts which have been greatly responsible for the dropping of the full-fledged wave theory of light are the phenomena of "photoelectric effects." If light waves hit the surface of zinc, this surface obtains a positive electric charge. Electrons (negative electric charges) of a mass m and the charge e are emitted by the zinc atoms. The absorbed energy E of the incoming radiation is used partly to overcome the attraction of the surface (work P) and partly to impart to the electron a kinetic energy $mv^2/2$. According to the law of the conservation of energy, this means $E = mv^2/2 + P$ or $mv^2/2 = E - P$.

If the wave theory of light were correct, the intensity of radiation would decrease inversely proportionate to the square of distance from the source of light. If we choose this distance great enough, the energy absorbed by a square inch of our zinc plate tends toward zero. However, experiments show that the kinetic energy $mv^2/2$ of the electrons which originates from the conversion of the incoming light energy into kinetic energy is not dependent at all on the distance of our zinc from the light source. The kinetic energy $mv^2/2$ and the speed v depend only upon the frequency (color) of the incoming radiation.

Einstein interpreted (1905) this result as follows: The light energy is not homogeneously distributed over the wave surface. Therefore, this energy is not thinned out at great distances, as the wave theory implies. On the contrary, according to Einstein's hypothesis, the light energy is concentrated in small packages of energy called "photons". To emit one electron, just one photon of light energy has to be absorbed by the zinc surface. The energy of a photon is proportional to the frequency $E = h\nu$, where h is a universal constant and ν the frequency of radiation. 'Frequency' means "number of vibrations per second". Therefore, the greater the frequency, the larger the photons and the greater the speed of the emitted photoelectric electrons. However, the greater the distance of the zinc from the source of light, the smaller the number of electrons which is emitted per second. But the speed of these electrons depends only upon the frequency of light and not upon the distance. This hypothesis is formulated mathematically by $E = h\nu$ and therefore $mv^2/2 = h\nu - P$. By checking this equation experimentally, one can determine the constant h. The first elaborate measurement was performed by Millikan. He found the value $h = 6.56 \times 10^{-27}$ erg. sec.

34. The Paths of Photons

If light consists of photons, we must assume that the brilliancy of illumination of a surface is proportional to the number of photons hitting a square inch. If we consider a phenomenon of interference or diffraction, the dark regions on a screen are regions where few photons hit the screen, while the diffraction maxima are regions where a great many photons come in. By this argument one understands easily that the motion of photons does not follow Newton's laws of mechanics but a very different type of law. One has to make use of the superposition of waves to find the distribution of photons. The single photon is not mentioned in these laws at all. The intensity of the light (the square of the wave amplitude) in a certain region is proportional to the number of photons in this region. 'The path of a photon' has no operational meaning. As a matter of fact, this path is not spoken of in any law of physics. We say: Our sun emits photons which heat and illuminate our earth. If we put a screen or a shutter in the way of these photons, we can predict the heat and luminosity effect of the radiation which hits or passes these devices. Since the laws of wave optics allow us by the superposition of waves to calculate the resultant amplitudes, we know in every region the average number of photons. But the laws of optics do not allow us to describe the path of a single photon on its way from the sun to the earth.

We can also understand more directly why 'the path of a photon' is an expression without operational meaning. If such a path could be physically produced, e.g., a rectilinear path, we could produce a light ray which travels along a geometrical straight line and passes through a definite point in a definite direction. However, the physical existence of such a light ray is incompatible with the laws of wave optics. In order to assure that a particular light ray passes through a definite point in space, we must make it pass through a very small hole in a screen. The smaller the hole, the more precisely the ray passes by a definite point. However, if the diameter of the hole becomes comparable in size with the wave-length of the light ray, the phenomenon of diffraction takes place. The ray passes through a definite point but does not continue in its original direction after passing the hole. If we want to have the ray continuing in its original direction, we have to make sure that the hole is much larger than the wave-length. But in this case the beam of light is of a considerable thickness and does not pass precisely through a definite point.

Therefore, if we want to produce a light ray by a physical operation, we cannot achieve an all-round approximation. Either we (1) bring it about that the ray passes approximately through a given point: then there will be no good approximation of direction; or we (2) achieve a good approximation in direction: then the precise location of the point is poorly defined. This state of affairs has been the basis of the "Principle of Complementarity" in physics.

We must have in mind that a "photon" cannot be thought of as a geometrical point. A package of light energy of the wave-lengths λ must contain a great number (e.g., N) of crests of waves in order to have the characteristics of a wave with the wave-length λ. Therefore, it must have at least the size $N\lambda$. If the photon is of a size which is small in comparison to $N\lambda$, it cannot have a distinct wave-length or frequency at all. 'The path of a photon' of the frequency $\nu = c/\lambda$ has, strictly speaking, no operational meaning, since only a "point" can traverse a "path".

We are accustomed to speak loosely about a light ray in empty space and to imagine it as a straight line. But here again this way of speaking has an operational meaning only if we assume some physical laws to be true. These laws are, roughly speaking, the independence of the light ray of its environment, in particular of the width of the opening through which it passes. But the validity of these laws means exactly the absence of the phenomena of diffraction, the absence of the wave properties of light. Since we know that these phenomena exist, 'path of a light ray', without including the environment of the light ray in the description, is an incomplete expression and has no operational meaning.

35. Laws of Motion of a Photon

We can easily give a quantitative description of this state of affairs. We may have a slit of the width A in a dark screen. A beam of light rays comes in perpendicularly to the screen and passes the slit. According to a simple calculation which we find in any elementary textbook of physics, only a fraction of the light energy moves behind the slit perpendicular to the screen. The balance is deflected. The bulk of it is deviated by an angle ψ (the first diffraction maximum) connected with A and the wave-length λ by the equation $A \sin \psi = \lambda$. If there were no diffraction, a light ray passing through a slit which has approximately the shape of a point ($A = 0$) would keep approximately its direction perpendicular to the screen. This means $\psi = 0$ approximately. But, because of the diffraction, A and ψ cannot be simultaneously zero or approximately zero if the wave-length λ is given. To produce an approximately rectilinear light ray, we have to make A and ψ both as small as possible. However, the smaller A becomes, the greater becomes ψ and vice versa. To achieve an approximately rectilinear light ray, we have therefore to compromise and to make both, A and $\sin \psi$, fairly small, e.g., $A = \sin \psi = \sqrt{\lambda}$. If we use light of a wave-length $\lambda = 10^{-5}$ cm. (ultraviolet), we would obtain $A = 0.003$ cm. and $\psi = 12$ angle minutes approximately. We can describe the light phenomenon by a rectilinear ray if we allow for a margin of 0.003 cm. in the starting-point and of 12 angle minutes in the direction. This means practically a straight line through a point and perpendicular to the screen. The smaller λ, the better is the approximation to a rectilinear path which can be made. If we use this "compromise" path instead of the theory of diffraction, we can predict within the margin described where our light ray will hit a screen put up behind the slit.

36. Mechanical Momentum of a Photon

If light radiation hits a material body, it exerts a force upon it—the pressure of light. This can be stated by saying that radiation imparts a certain momentum upon this body. We can interpret this fact by ascribing to every photon a mechanical momentum exactly as to a moving mass. The action of radiation to a body can be treated as collision between masses. Each mass m has a momentum mv, if the speed is v. The sum of the momenta is not changed by the collision, according to Newton's Third Law of motion. If one of the bodies loses momentum, the other one must gain it. A photon has an energy $E = h\nu$. According to the theory of the electromagnetic field, a package of radiant energy E carries a momentum $M = E/c$. This means: If radiant energy E

hits a material body, it imparts the same momentum as if a mass m with the speed v hit the body, mv being determined by $mv = E/c$. A photon in particular has the momentum $M = h\nu/c$. Since every photon travels with the speed of light c, this momentum does not depend upon speed but on the frequency or wave-length. Since $\nu = c/\lambda$, we have

$$M = \frac{h\nu}{c} = \frac{h}{\lambda}.$$

This momentum can be directly checked by the Compton effect, which we observe when we examine the collision of a photon with an electron. By this collision the momentum of the electron as well as that of the photon is affected. The alteration of the momentum of the photon means an alteration of the wave-length. If the electron has originally the momentum of zero, the momentum of the photon can only decrease by the collision. This means that the wave-length λ must increase and that the frequency ν must decrease. From the hypothesis that the momentum of the photon is $M = h/\lambda$, one can derive that the wavelength of a photon which is moving after the collision at a right angle to its original direction increases by h/mc, where m is the mass of the electron. $h/mc = 2.42 \times 10^{-10}$ cm. is called the 'Compton wave-length'. This alteration of wave-length can, of course, be observed only if the original wave-length is very small. Compton carried out this experiment by using X-rays the photons of which have very short wave-lengths. He could confirm the alteration of the wave-lengths of X-rays which hit some material bodies which are rich in electrons. This experiment is the most direct measurement of the mechanical momentum h/λ of the photon.

37. The Interpretation of Diffraction in Terms of the Mechanics of the Photon

If we keep in mind that the photon carries a mechanical momentum, we can give to the diffraction experiment also a purely mechanical interpretation. Instead of attempting to produce an exactly rectilinear light ray, we can attempt to produce a photon which has a certain definite position and a momentum in a definite direction. We determine the position again by having the photon pass a slit of a size A in a screen. If no diffraction existed, the whole momentum would be perpendicular to the screen; the momentum parallel to the screen would be zero. By decreasing A more and more, we could achieve with an arbitrary accuracy that the photon passes through a given point and has a momentum exactly perpendicular to the screen. But, because of the diffraction, the momentum of the photon has also a component parallel to the screen. Since the momentum itself has the magnitude h/λ, its component parallel to the screen is $h/\lambda \sin \psi$. We denote this component by 'p'. From $A \sin \psi = \lambda$ (sec. 35) follows $A \sin \psi = h/v \sin \psi$ or $Ap = h$. We see again that it is impossible to produce a photon which has simultaneously an exact

position and an exact direction of momentum (perpendicular to the screen). For in this ideal case we would have $A = 0$ and $p = 0$, which would render $Ap = h = 0$, which is impossible, since h is different from 0. To achieve an exact position ($A = 0$), we use an infinitely small slit. In this case p would increase infinitely. We could not say anything about the direction of the momentum. If we want an exact direction of momentum (p very small), we need a very great A. This means that the position of the photon becomes undefined. We can again compromise by making A and p both fairly small. Then we have a photon with a fairly exact position and fairly exact direction of momentum.

We understand very clearly by these considerations that the photon is not a particle with a certain position and momentum which both exist but cannot be measured simultaneously. Actually, an experimental arrangement which produces a photon with an exact position (very small A) frustrates the achievement of an exactly perpendicular momentum (p very small). 'A photon with an exact position and an exact momentum' is an expression which does not enter into any description of physical facts; it does not enter into the formulation of any physical law. Therefore, this expression has no operational meaning and has to be dropped from the vocabulary of physics.

If we assume the x-axis parallel to the screen, we may introduce the symbols $A = \Delta x$, $p = \Delta p_x$, where Δx means the inexactitude of the position and Δp_x the inexactitude of the momentum, both in the x-direction. If we describe a state of affairs by saying that our photon has a position $x = 0$ and the momentum $p_x = 0$ parallel to the screen, this description is only correct if we allow the "uncertainty" Δx in the position and Δp_x in the momentum. Since $Ap = h$, we have the relation $\Delta x \times \Delta p_x = h$ between these "uncertainties".

If we use the description $x = 0$, $p_x = 0$ for the present state of the photon, we can predict the future with the "indeterminacy" Δx and Δp_x by assuming a rectilinear path perpendicular to the screen. These remarks are the roots of W. Heisenberg's relation of "uncertainty" or "indeterminacy" (sec. 43).

VII. Mechanics of Small Masses (Wave Mechanics)

38. Failure of Newton's Mechanics in the Domain of Very Small Particles

Newton's physics assumed that the laws which govern the motion of the planets govern also the motions of the smallest particles of matter—the atoms and electrons. One concluded that the orbits of the electrons revolving around the atomic nucleus would follow the same law as the orbits of the planets around the sun, except that the gravitational forces in the solar system had to be replaced by electrostatic attractions in the atom. However, both types of forces followed one and the same law: The force is inversely proportional to the square of the distance. But in quite a few cases, and in crucial ones at that, the Newtonian pattern failed to render an adequate method of describing and predicting the conditions of atomic equilibrium and motion.

In spite of the general belief that the action of chemical valences can be described by the Newtonian concept of force, it has been actually impossible to substitute into the Newtonian pattern '$ma = f$' an expression for 'f' which would describe the phenomenon of saturation of a valence. It is easy to understand that two particles of unlike electric charges (positive and negative ions) can attract each other. But how can two neutral particles form a compound? However, just this seems to happen in the case of very simple chemical compounds (homopolar compounds). The hydrogen molecule H_2 consists of two neutral hydrogen atoms. The chemist says that the valences of these atoms saturate each other. But from the point of view of the mechanics of electrostatic forces it is hard to understand how this can happen and how a third hydrogen atom can be prevented from being attracted. The phenomenon of saturation of valences remains unexplained in Newton's mechanics.

Still more obvious is the failure of Newton's pattern to describe the motion of electrons in the atoms which must occur in order to explain the emission of the spectral lines. According to the theory of Niels Bohr, the emission of spectral lines from heated hydrogen gas is a kind of reversed photoelectric effect. If a photon hits the surface of a metal, the radiant energy of the photon is converted into the kinetic energy of an electron leaving the surface (sec. 33). Similarly, if the energy (kinetic and potential) of the moving electrons within the atom decreases, the mechanical energy lost by the atom is converted into the radiant energy of the photons which leave the atom. These photons constitute light of a specific wave-length which reveals itself by the spectral lines of the hydrogen gas.

39. Bohr's Theory of the Emission of Light by the Atom

When the atom passes from the energy E_n to the smaller energy E_1, according to Bohr's hypothesis exactly one photon of the energy $h\nu$ is produced. If we accept Einstein's fundamental hypothesis about the conversion of mechanical into radiant energy and vice versa, we have the equation $E_n - E_1 = h\nu$. If we know the initial and final energy of the atom (E_1 and E_n), the frequency ν of the emitted radiation is determined. We know from the examination of the frequencies of spectral lines that there are specific frequencies in the hydrogen spectrum. Therefore, there can be only specific values of energy in the hydrogen atom.

If we consider the simplest case, we can assume that the negatively charged electron is moving along a circle around the positively charged nucleus. According to Newton's laws of motion and Coulomb's law of electrostatic forces, the attraction e^2/r^2 must be balanced by the centrifugal force mv^2/r, where e is the charge of the electron and of the nucleus, m the mass of the electron, v the linear speed, and r the radius of the circular orbit. From $e^2/r^2 = mv^2/r$ it follows that $v^2 r = e^2/m$. This means that for every radius r a circular motion is possible if we choose only the speed v accordingly. The value of the energy E of our circular orbit is given by

$$E = - e^2/r \text{ (potential energy)} + mv^2/2 \text{ (kinetic energy)} = - e^2/2r\,.$$

Since there is an orbit for any value of r, there is also an orbit for any value of energy E. Therefore, Newton's mechanics cannot help us to derive the specific values of energy (E_n and E_1) which render specific values of frequency. Bohr, following some suggestions of Planck's theory of radiation, assumed that the orbits of electrons have to obey, besides Newton's laws of motion, a still different kind of law, called "quantum laws", by which most of the orbits allowed by Newton's mechanics are excluded.

These quantum laws require that the angular momentum of the revolving electron has to be equal to $h/2\pi$ (where h is the same constant as in secs. 33–37) or an integer multiple of $h/2\pi$. This means that mrv is equal to $h/2\pi$ or $nh/2\pi$, where n is an integer number. Then it follows that the smallest possible value of the radius r can be derived from $2\pi mrv = h$ and $mv^2r = e^2$. By eliminating v, it follows that $4\pi^2m^2r^2v^2 = h^2$, $4v^2mr = h^2/e^2$, and, finally, $r = h^2/4\pi^2me^2$. This value of r provides us with the smallest possible size of an electronic orbit. The charge e of the electron is known from Millikan's experiment on the ionization of oil droplets; m is known from J. J. Thomson's experiments on the deflection of cathode rays in magnetic and electric fields; h is known from the photoelectric effect (sec. 33). From the known values of h, m, and e we obtain that r is approximately 10^{-8} cm. This value agrees with the size of the atoms obtained by other methods, e.g., Brownian motion, diffraction of X-rays by crystal lattices, etc.

Knowing r, we can calculate the energy E_1 of the smallest orbit ($n = 1$). We obtain $E_1 = -2\pi^2e^4m/h^2$. To obtain the energy of the nth orbit, we have to replace h by nh and obtain $E_n = -2\pi^2e^4m/h^2n^2$. This formula allows us, by using the Einstein-Bohr formula $E_n - E_1 = h\nu$, to calculate the frequencies of the spectral lines of the hydrogen atom in agreement with the experiments.

40. The Wave Theory of Matter

This was a great success of Bohr's spectral theory, but we must not forget that it was achieved by a partial abandonment of Newton's mechanics without replacing it by a new mechanics. The crucial formula, $4\pi^2m^2r^2v^2 = h^2$, was obtained by a hypothesis (quantum law) which was superimposed upon Newton's laws of motion, leaving these laws themselves substantially unaltered. In this way mechanics and optics became an incoherent patchwork which gave rise to many pseudo-problems.

The French physicist, Louis de Broglie, looked at this situation from a new point of view. He did not believe in patching up Newton's laws by the quantum laws but in altering Newton's laws themselves. He took his cue from a comparison between optics and mechanics. The optical phenomena can be described and presented in terms of light rays (straight lines in vacuum) in every case where diffraction can be neglected. If diffraction comes in, we have to pass from ray optics (geometrical optics) to wave optics. Under what conditions must diffraction be considered? Obviously, diffraction comes into the picture if light passes through small openings or around small obstacles, the word 'small' meaning 'comparable in size with the wave-lengths of light'. This means that the pure ray optics loses its applicability if the rays have a great curvature. The word 'great' means here again a radius of curvature which is comparable with the wave-lengths of light. Perhaps, argued De Broglie, the orbits calculated from Newton's mechanics play only the role of the light rays in optics. Perhaps Newton's mechanics loses its applicability also if the orbits have a very large curvature or a very small radius of curvature. This is probably the case if we consider the curvature of the electronic orbits around the nucleus of the hydrogen atom. The radius of curvature of the smallest orbit is of the order of magnitude of 10^{-8} cm., much smaller than the wave-lengths of visible light, which is about 10^{-5} cm.

Under such circumstances, De Broglie argued, we have to replace Newton's mechanics by a new mechanics. This generalized mechanics would be in such a relation to Newton's mechanics as wave optics is to ray optics. For small curvature we can regard the optical rays as paths of the photon. But, if diffraction comes in, the behavior of photons can no longer be described by paths (secs. 34, 35).

Louis de Broglie advanced the idea that there is a new type of waves called by him "waves of matter" and later, after him, "Broglie waves" which determine the behavior of material particles and electrons in a way similar to the way in which the light waves determine the behavior of photons. The most natural hypothesis was to assume that the relation between wave-length and mechanical momentum is for the material particle the same as for the photon. We know that for photons the momentum p obeys the formula $p = h/\lambda$. If we know the λ of the kind of light used, we can calculate from this formula the momentum p. In the case of material particles, however, the momentum is known from mechanics $p = mv$ (v being the speed of the particle), while the wave-lengths of the De Broglie waves must be calculated from the formula $\lambda = h/p = h/mv$.

41. The New Laws of Motion for Small Particles

One notices immediately that if the mass m of a particle tends toward infinity, the Broglie wave-lengths λ tend toward zero. This means that the phenomena of diffraction can be neglected. The motion of great masses can be treated without using the waves. We have to do with orbits of masses which obey Newton's laws of motion. But if we have to do with very small particles, the Broglie wave-lengths become comparable with the radius of curvature of the orbit, as in the case of the electronic orbit in the hydrogen atom, and we have to apply the laws of "wave mechanics". According to these laws the term 'path of a particle' has no more operational meaning than 'path of a photon' in ordinary optics. What we describe according to Newtonian mechanics as a stream of electrons emitted by some source (heated wire or the electrodes of a vacuum tube) must now be described as the emission of "Broglie waves". Exactly as the intensity of the electromagnetic waves describes the number of photons per unit of volume in a certain region, the intensity of the De Broglie waves describes the number of material particles per unit of volume—the density of mass.

The first confirmation of this daring hypothesis was De Broglie's derivation of Bohr's "atomic radius" without using the quantum laws. The wave propagation along the circular orbit of the electron around the nucleus must not destroy itself by superposition and interference. Therefore, the perimeter of the smallest orbit must be equal to one wave-length. This means, according to De Broglie, $2\pi r = \lambda = h/mv$. This result is obviously equivalent to Bohr's quantum law $mrv = h/2\pi$ (in sec. 39).

42. Diffraction of Material Particles

The most direct confirmation of De Broglie's hypothesis, however, would be to show that material particles (e.g., electrons) which pass a small hole produce on a screen a diffraction pattern similar to the diffraction pattern of light. There would be alternatively dark and bright rings. The bright rings are the regions of the screen where a great many particles hit, while the dark rings show us the regions where no, or very few, particles hit. The size of the hole must be comparable to the wave-lengths of the De Broglie waves. The American physicists Davison and Germer used the intervals between the atoms of a metallic foil as slits to produce a diffraction pattern by having electrons pass this foil.

By these and similar experiments it has become evident that very small material particles follow laws of motion which are very different from the laws of Newtonian mechanics. The particles of a beam which pass through the hole in a direction perpendicular to

the screen do not continue moving in this direction behind the screen as they are expected to do according to Newton's law of inertia. Only a certain percentage of the incoming particles do so. They hit the center of the receiving screen and form the central maximum of diffraction. The greatest percentage of the remaining particles is diffracted toward the first diffraction maximum by the diffraction angle ψ, where $A \sin \psi = \lambda$. If we know that, e.g., a thousand particles enter the hole, we cannot predict what every individual particle will do behind the slit. But we can safely predict that a certain percentage will hit the central maximum and a certain percentage the first diffraction maximum ψ, etc. Here again, as in the case of the photon, we can denote the size of the slit by Δx (this means $A = \Delta x$) and the momentum of the particle parallel to the screen by Δp_x. Since the formulas are the same as in the case of the photon, we have again $\Delta x \times \Delta p_x = h$. The product: the "uncertainty of x" times the "uncertainty of p_x" equals a constant h.

43. The Uncertainty Relation

The meaning of this relation in the case of material particles seems to be much more paradoxical than in the case of the photon. For this relation is incompatible with the basic statement of Newtonian mechanics that every mass point has at every moment of time a definite coordinate x and a definite momentum $p_x = mv_x$ in the direction of x. If we know these values for an instant of time $t = 0$, Newton's laws of motion allow us to predict these values for any future instant of time. If we examine carefully our result obtained in the case of photons and apply it to the case of small material particles, we learn that there is no law of mechanics which contains the expression 'position and momentum of a particle at a certain instant of time'. There are experimental arrangements which we can describe by saying that the particle passes the diaphragm at a position $x = 0$ with a margin of error Δx; by narrowing the slit, we can make this margin Δx as small as we wish. But then the equation $\Delta x \Delta p_x = h$ shows us that Δp_x becomes very great. This means that we only know that behind the slit the momentum p_x of the particle is zero with the allowance of a wide margin of error. If we consider a great number of particles passing the slit of the width Δx, most of them have a momentum $p_x = h/\Delta x$ parallel to the screen. But also all smaller and greater momenta will occur with a certain frequency according to the Gaussian distribution of errors. If we fix the position of the diaphragm and of the small hole at $x = 0$ relative to the earth, we obtain diffraction rings on a screen which is at rest behind the slit. We can predict the position of these rings exactly from knowing only the position of the screen relative to the slit. Our information which allows us to predict the behavior of the particles behind the slit is based entirely upon the measurements (in the ordinary sense of the word) which we

can perform in the experimental set-up. In this way we know $A = \Delta x$ (width of the slit) and the distance of the screen from the slit. This information allows us to predict the position of the diffraction maxima on the screen. The momentum Δp_x of an individual electron while passing the slit remains unpredictable. But obviously it is possible to "measure" the momentum of an individual electron in a certain sense. Since the component Δp_x of this momentum is directed parallel to the diaphragm, the particle will, in passing the slit, impart a momentum upon the diaphragm parallel to the diaphragm itself. If the diaphragm is not fixed with respect to the earth, but movable, one can find the momentum of the particle by measuring the momentum of the diaphragm, which means practically its speed obtained by the impact of the particle. But if the screen is moving while the electrons pass the slit, the position of the particle relative to the earth is not predictable even if the slit Δx is very small.

We must always decide in what result we are interested: in the position of the particle passing the diaphragm, or in the momentum of this particle. In each case we can make a prediction. In the first case we can predict the diffraction pattern on the screen. In the second case we know the momentum which the diaphragm gets from the particle, and we can make predictions by means of Newton's mechanics. We can predict the motion of bodies which are hit by the diaphragm. We could, e.g., predict the elongation of a ballistic pendulum.

44. Misunderstandings about the Relation of Uncertainty

We must carefully avoid the misunderstanding which has been caused by the way some physicists have discussed the relation of uncertainty. One hears sometimes the statement: "It is impossible to measure simultaneously the position and the momentum of a small particle". This sounds as if there would be small particles which possess certain positions and certain momenta. We are told that we can measure either of them but that nature is so diabolic as to prevent us from measuring both simultaneously. This statement is rather misleading. The expression 'a particle with a certain position and a certain momentum' has no operational meaning if De Broglie's hypothesis is accepted. However, we can set up an experimental arrangement which gives to the expression 'particle with a certain position' a kind of approximate operational meaning. We can make some predictions from such statements. They determine the future within a certain margin of "indeterminacy". There are also arrangements which allow us the use of the expression 'particles with a certain momentum'. They lead to predictions which are also reliable within a certain margin. The relation of "uncertainty" involves the relation between these two margins of prediction and is called, for this reason, the "relation of indeterminacy". But "a particle with a certain position and a certain momentum" is not an object of the physical reality as far as very small particles are concerned. Speaking exactly, a particle by itself without the description of the whole experimental setup is not a physical reality. But "a particle passing a diaphragm at a certain point" is a description of a physical reality and "a particle imparting to a diaphragm a certain momentum" is a description of a physical reality too.

As a matter of fact, we would find all these statements less paradoxical if we dropped the word 'particle' altogether from the language of physics. We could say that the mechanics of the smallest particles, wave mechanics, allows us from the knowledge of the initial experimental conditions to predict the future observable phenomena. Among these phenomena are in particular "point events", e.g., scintillation produced at a point of a screen, and "impulse events", e.g., a momentum imparted to an observable material body. Then nobody would be puzzled by learning that scintillations on a screen do not follow an orbit which is determined by Newton's laws of motion. Hans Reichenbach pointed out that small particles with definite positions and definite momenta can be introduced into the language of physics and that their introduction can be justified in a certain sense. The logical coherence of the system of physics would not be violated, but the motions of these particles would follow very awkward laws. Reichenbach speaks of "causal anomalies". He calls this way of describing the subatomic phenomena the "exhaustive interpretation", while he calls the presentation in Bohr's complementary language as in the present monograph the "restrictive interpretation". The latter sticks strictly to experience and to simple operational definitions but sacrifices the traditional language of mechanics. The exhaustive description sticks strictly to the language of Newtonian mechanics but requires very involved operational definitions. It gives a certain satisfaction to the lover of logic but does not give the simplest possible description of the subatomic phenomena.

45. Bohr's Idea of Complementarity

As Niels Bohr has pointed out repeatedly, the physicist feels at ease when he can keep to the language of the mechanics of everyday life as far as possible. However, if we stick to "simple and practical" operational definitions there is no adequate means of describing the subatomic phenomena in terms of "full-fledged particles" with position and momentum. In this situation the physicist takes some satisfaction in speaking, at least, of "particles which have only position" and of "particles which have only momentum". In this way we use, at least partially, the language of corpuscles to which we are accustomed in Newton's mechanics.

If we admit any operational definition without regard to whether it is "simple and practical", we can, of course, measure the exact position and velocity of a particle and give an "exhaustive interpretation" of its state. We use two diaphragms with small slits and observe a particle which passes both slits. Then the velocity can be calculated from the distance of the slits. But the velocity calculated in this way is not the velocity which helps us to predict the future course of the particle. For after the passing of the second diaphragm we must again take into account diffraction, and the future course cannot at all be calculated from the velocity. The "exhaustive interpretation" of the state of a particle is no basis for predicting its future states by simple laws. Therefore, the correct statement of the "relation of uncertainty" is not that "position and velocity cannot be measured simultaneously" but that "there is no law of prediction which contains reference to the simultaneous position and velocity of a particle". If a statement does not contain a rule of prediction, it is not a physical law but purely tautological. It is a definition. Our statement about measuring velocity by using two diaphragms is certainly an operational definition of position and velocity. But this definition does not enter into any law of physics which would help us to predict the future course of events. As we emphasized in section 2, an operational definition is only "helpful" or "practical" if it helps us to formulate physical laws.

The position and the momentum of a particle are two physical properties which reveal themselves on different occasions which exclude each other. The position can be obviously observed as a "point event" and the momentum as an "impulse event"—they have nothing to do with each other. Only if the particle has such a great mass that Newton's mechanics can be applied, can we speak of a particle having a position and momentum and can we calculate the momentum from two positions by the formula $p = mdx/dt$. According to Bohr, the physical world cannot be described by one coherent language. There are two languages which "complement" each other. Under certain circumstances the language of "positions of particles" or "point events" must be used; in other circumstances, excluding the ones above mentioned, we speak of "momenta of particles" or "impulse events". If we make use of all possible information about the present state of the world, we must use both languages. Then we can predict all events which our actual science enables us to predict. This aspect of the world is what Bohr has called the "aspect of complementarity".[25]

The indeterminacy relation $\Delta x \Delta p_x = h$ can also be written $\Delta x \cdot \Delta v_x = h/m$ if we introduce $p = mv_x$. If the mass m increases, h/m tends toward zero. It becomes more and more meaningful to assume that Δx as well as Δv_x are 0 and to speak of a mass with a definite x and a definite v_x.

We understand now very well what prediction can be made if we send particles through a very small slit in a diaphragm. We consider, e.g., a beam of electrons directed perpendicularly to the diaphragm. We proceed as we did in optics (sec. 35) in order to obtain a "compromise beam". This means to make the size of the slit rather small but not so small as to make the momentum undefined. Then we obtain what we call in experimental physics a "beam". According to Newton's mechanics, it would follow from the law of inertia that all electrons, after passing the slit, continue in their course perpendicular to the diaphragm. They strike a screen which is set up parallel to our diaphragm within a region which is nearly congruent to the slit. However, if diffraction plays a certain role, there is for every region of the screen a certain possibility of being hit by a particle. Instead of the rigid law that one limited region is hit and all the rest of the screen remains untouched, we have to say that the frequency of a hit is distributed over the whole screen. One says sometimes, and it means the same thing, that the law of inertia is replaced by a law predicting the statistical distribution of hits. We can only say that in the region where, according to Newton, all particles should hit there is merely an overwhelmingly great frequency of hits.

46. By What Law Does Wave Mechanics Replace Newton's Second Law?

We assume now that the beam of electrons passes a "field of force". We consider only the simplest case where the force has the direction of velocity. The force f is defined by the formula $f = ma$, where 'a' means the acceleration of a "great mass" m to which Newton's law can be applied. We assume, e.g., that this beam is launched upward against the force of gravity. The acceleration of gravity at this point of the earth may be g. The particles may start with an initial speed v. According to classical mechanics, they would move upward by a distance s which is given by the equation $mv^2/2 = mgs$ or $s = v^2/2g$ (decrease of kinetic energy is equal to the work done against the force of gravity). If the mass m is "small", Newton's mechanics can no longer be applied. By "small" we mean that there is a considerable drop of potential between two points with a distance of one "Broglie wave length": $\lambda = h/mv$. We find now from the laws of wave mechanics that we can affirm only that most of the particles reach the height of $s = v^2/2g$. But there is for every height s a certain percentage of particles which reach it. Newton's law of motion ($ma = f$, where f is the force of gravity) is replaced by a formula which gives us the percentage of particles which reach a certain height s.

This reshaped law of motion becomes particularly important if we replace the force of gravity mg by an electric field which acts upon the charge of the electron. If the potential difference between the diaphragm and the final position of the electron is V (voltage), the force in the Newtonian sense is eV/s. If the mass m is so great

that Newton's mechanics can be applied, the distance s which an electron traverses is again given by $mv^2/2 = eVs$, and therefore $s = v^2m/2Ve$. If the mass is "small", we can no longer predict that all particles will reach s but say only that most of the particles reach the distance indicated by the last formula. However, some particles go farther. There is for every distance s a certain percentage of electrons which reach it.

This result of the new mechanics turned out to be particularly important for the understanding of the phenomena of radioactivity. The alpha radiation of radioactive substances (like radium) consists in the emission of positively charged particles (alpha particles) from the nucleus of the atom. However, if we measure the speed v of the emitted particles, we find that, according to Newton's laws of mechanics, they must have traversed a considerable electrical potential difference in order to acquire this speed. But in this case the nucleus must be surrounded by a very high "potential barrier". This means that there is also a considerable potential difference between the center of the nucleus and its surface. But then it can be calculated from Newton's laws and the observed speed v that the charged alpha particles will not succeed in "ascending" this potential difference and in leaving the atom, as our v would be too small. The Russian physicist Gamow pointed out that, according to wave mechanics there is for any initial value of v a certain percentage of particles which will succeed in reaching the surface of the nucleus and be emitted into the open. In this way the new mechanics finds a wide field of application in the motion of the subatomic particles. According to the new mechanics of small particles, we can start from the observable initial conditions, e.g., the initial electron beam, and predict future observable phenomena. We are not able to predict every single point event, but, if the "force" is given, we are able to predict the statistical distribution of the future point events.[26]

47. Physical Reality and Causality

One must not exaggerate the gap between the new mechanics of small masses and the Newtonian mechanics. According to Newton's mechanics, the future positions and velocities of mass points can be predicted with certainty. This is, of course, only true if we speak in the language of the "calculus" (sec. 1), the language of "physical quantities". However, if we substitute for these quantities the operational meaning, some statistical component always enters. Speaking in terms of sense data, the distinction between prediction of a single event and statistical prediction loses even its clear meaning. For the question always remains arbitrary to a certain degree of what we are to regard as a single sense datum and what as an average of a great many sense data. The difference between a statistical theory and a strictly causal theory is not as much in the realm of sense data as in the realm of physical quantities.

Newton's mechanics allows a strictly causal prediction of the positions and velocities of mass points which are regarded as the physical reality described by this theory. But what is the physical reality described by the new mechanics? If we regard the percentage of particles producing point events in a certain region (the wave function) as this reality, the new theory is causal too. It only loses its causal character if we regard the positions and momenta of particles as the state variables which describe the physical reality. But these variables are in the new mechanics not state variables in the sense in which this word is used in Newtonian physics. There is no description of the world at a certain instant of time which contains both position and momentum of a mass point.

Moreover, while in Newton's mechanics the state of the mass point is described without an experimental setup to which it belongs such a "mass point stripped of its environment" does not occur at all in the new description of the world. This new description which contains, e.g., the term 'position of a mass point' gives a description of the present state of the world by describing actually objects like "a mass point in a particular experimental setup." These objects have, obviously, operational meaning. And there is a "complementary" description of a second setup which allows us to ascribe a "momentum" to the particles. But there is no setup which would allow us to use both the terms 'position' and 'momentum of a particle'. Obviously, we can make only predictions, starting from statements which have operational meaning. Therefore, we can make predictions starting from a statement about the "position of a particle". But these predictions are not so precise as the predictions in Newton's physics which start from statements about both "position and momentum of a particle". To say, therefore, that in wave mechanics the law of causality is not as strictly valid as in Newton's mechanics would be a very inadequate description of the character of the new mechanics. For these two systems of mechanics do not use the same state variables to describe the physical reality. Therefore, we cannot compare them as to the greater or smaller validity of the law of causality. Such a comparison would make sense only if the old and the new description made use of one and the same set of state variables. But this could be achieved only by reducing both descriptions to the language of sense data or to the "thing-language". But in this language the distinction between a "causal" and a "statistical" theory becomes vague.

The new mechanics, we are often told, does not describe physical reality at all. For the state of a photon or of a mass particle is never described objectively but always in connection with an experimental setup or an instrument of measurement with which it is in interaction. We cannot, e.g., describe the state of a photon on its way from the sun. We can describe it only at the instant of time when it strikes a screen or comes otherwise in interaction with a material body which has so great a mass that its behavior can be described by Newtonian mechanics. Even a revolutionary mind like Einstein's has for a long time doubted whether it is advisable to drop the traditional conception of objective physical reality, according to which "physical reality" is ascribed to a material particle without regard to its environment. Einstein has suggested not to stop efforts to find a theory of subatomic phenomena which describes "real things" in the traditional sense. Obviously, there is no convincing reason which could prevent us from sticking to the Newtonian concept of physical reality and from regarding Bohr's "Principle of Complementarity" as a provisional state of science. But one may as well admit a different type of world description in which the expression 'position of a particle' cannot be applied without describing the instrument by which this position is measured.[27] If this description is "simple and practical", we can as well ascribe "physical reality" to the objects of our new mechanics, provided we mean "reality" in the operational and not the metaphysical sense.

If we examine the whole question of causality more thoroughly, we soon notice that this law in its whole generality cannot be stated exactly if the state variables by which the world is described are not mentioned specifically. Otherwise, the general formulation of the law of causality would have no operational meaning. For the more general formulation says: let a state A of the world be succeeded by a second state B. If A occurs again, B follows again too. If no specific state variables are used for the description, we cannot check whether the state A has recurred. Moreover, only if we know that certain physical laws are valid can an operational meaning be assigned to the state variables (sec. 3). Therefore, the type of physical laws which are to be set up is dependent upon the kind of state variables which are used.

48. Object and Subject in Wave Mechanics

Frequently we are told that in the new physics the role of the perceiving subject is greater than in Newtonian mechanics. The observing physicist has to be introduced explicitly into physics, while previously physics has dealt only with the observed object.

This assertion must be taken with a grain of salt. We have seen that in wave mechanics the state of a mass point can never be described without including the instrument of measurement which is used. We must state explicitly whether we use a narrow slit or a wide slit in a diaphragm in order to describe the position of a mass point. If we call the mass point the physical object which we wish to describe, we must include into the description the instrument of measurement. This instrument is a medium-sized mass and can be described by using the language of Newton's mechanics. But there is

no sense in calling this instrument of measurement the "observing subject."

If we use the expression "observing subject" as it is used in the psychology of everyday life, we have to say that the observing subject (the living physicist) observes the instrument of measurement in the same way as he observes any object of Newtonian physics. Therefore, by "observed objects" we have to understand in the new mechanics, as well as in Newton's mechanics, the medium-sized mass, the instrument of measurement, the scale or balance which one observes immediately. The electron which passes a diaphragm must not be called the "observed object" if we want to avoid ambiguities. The "electron" is a set of physical quantities which we introduce to state a system of principles from which we can logically derive the pointer readings on the instruments of measurement. If we call these instruments "physical objects", the "electron" or "photon" can be called so only by a certain shift in the sense of the term 'physical object'. But we must never forget that there is no meaning in the question of what is "really" a "physical object". The only problem is to agree about an unambiguous use of this word. The loose way of speaking which is so customary in the borderland between physics and psychology has even succeeded in confusing the instrument of measurement with the "perceiving subject".

To say that the observer destroys by the very act of observation some property of the object which he wants to observe is a misleading formulation too. The observer has nothing to do at all in this matter. His role is exactly the same as in Newtonian physics. The "momentum of a particle" cannot be destroyed by observing the position because this momentum has never existed except in so far as we have a setup which allows the definition of a "momentum." All this sensational "destruction by observation" is an oversimplification. We found a similar situation in the theory of relativity. The introduction of the observing subject was misleading too. The correct thing to say is that, by setting up an experimental arrangement which would allow us to define 'the position of a particle', we do not "destroy the momentum of this particle". What is "destroyed" is only the possibility of setting up the "complementary" arrangement, which would allow us to define the term 'momentum'. The role of the living observer is also in "Relativity" only to observe "clocks" or "yardsticks" which can be described in the language of everyday life.

49. Metaphysical Interpretations of Wave Mechanics

Quite a few authors have maintained that by the new mechanics an "irrational element is introduced into physics" or that "physics is now supporting an idealistic word picture" or that "physics is now in agreement with the doctrine of free will" or even that "by the wave mechanics for the first time in the history of human thought the conflict between religion and science has been settled." If we pursue precisely the analysis of the logical structure of the new mechanics, we will understand that there is no foundation for all these philosophical interpretations of the new mechanics.

The basis of these interpretations has been provided by some metaphysical formulations of the new mechanics which have been given by a great many philosophers and, for that matter, by quite a few physicists and mathematicians. Such a formulation is, e.g., the introduction of the observing subject as a mental, immaterial "entity" into physics itself. This way of speaking, as we have already seen, is inadequate. By others the new mechanics is described as introducing a physical object which is both particle and wave. Some authors speak of it as "particle and/or wave" and some authors have even coined new words like "wavicle" to denote this hybrid object. As a matter of fact, such an object resembling a centaur, who was half-man and half-horse, does not exist in the new mechanics. There are experimental setups which can be described by using the term 'position of a particle' and others which can be described by using the terms 'momentum' or 'wave-lengths'. All the confusion is produced by speaking of an object instead of the way in which some words are used. We have again an example of what Carnap calls the fallacy of using the "material mode" of speaking instead of the "formal mode". In a great many presentations of the new mechanics it is not even pointed out clearly that the so-called "dualism between particle and wave" is only another expression for the complementary use of the words 'position' and 'momentum of a particle'. As in a great many other cases, an ambiguous way of speaking is introduced by the predilection of quite a few authors for an "ontological" way of speaking. Some scientists who are very competent in their fields succumb to the temptation to make statements on ontology. This is quite natural since ontology is nothing but the use of our everyday language in a domain where it loses its meaning.

The mental or idealistic character of the new mechanics is occasionally demonstrated by calling the De Broglie waves "waves of probability". Since "probability" is often used as a psychological concept, we describe in wave mechanics the physical world by using terms of psychology. This interpretation is certainly a misleading one. The new mechanics describes the percentage of electrons which strike on the average a certain region of the screen. There is nothing psychological involved, or at least there is no more psychology than in any branch of physics. For even in Newtonian physics every statement is accompanied by its operational definition. And this definition contains the description of physical instruments. This description again makes use of words which express sense data like 'blue', 'warm', 'soft', etc. These words are the only

psychological element in modern as well as in traditional physics.

Another way in which metaphysical misinterpretation enters into physics is the use which is made of the word 'real' in contrast to the word 'fictitious'. Quite a few authors would say that the De Broglie waves are "not real waves" but only a "fiction" which serves to describe the path of particles. This way of speaking is also misleading and is not in agreement with a coherent presentation of physics. We have learned that the movement of photons is determined by the electromagnetic waves exactly in the same way as a movement of small particles (electrons) is determined by the De Broglie waves. The reality of the electromagnetic waves can be demonstrated only by checking the effect of the photons on the bodies with which the waves come in contact. In the same way the reality of the De Broglie waves can be demonstrated too.

Quite a few philosophers and scientists have claimed that the "Indeterminacy Relation" gives a support to the philosophical doctrine of "Free Will". However, no statement can be supported by physical theories which cannot be formulated in terms of physical operations (sec. 2). The statement "the will is free" has, certainly, no operational meaning. It is a purely metaphysical statement and cannot be supported by any physical theory (secs. 2 and 4). Therefore, all the talk about the intrusion of mental and psychological elements into physics has its source only in an inadequate presentation of the recent parts of physics.

VIII. Structure of Matter

50. Continuity or Discontinuity of Matter

It has been an age-old dispute whether matter tightly fills the world space or whether matter consists of small indivisible particles, "atoms", between which there is empty space. The existence of these atoms has been suggested by many facts, in particular the law of constant proportions in the chemical compounds, the elastic properties of gases, etc. However, it has been held out against the old atomistic theories that nothing can be won by assuming that a big piece of iron consists of very small pieces of iron, called atoms. For then the question has to be raised: What is the structure of the small piece of iron called atom? If it is iron, it must have the same structure as the big piece, and so on.

The physics of the twentieth century has come to the firm conviction that the mass density inside solid bodies shows conspicuous maxima and minima which have specific distances from each other. These distances are characteristics of the physical and chemical constitution of the body. They are usually interpreted as the intervals between the centers of atoms. But it would not be correct to say that in these intervals between the atoms the space is "empty" and that the density of matter has the exact value zero. As the "position of a particle" is not an element of "physical reality", statements like "This volume of space is empty" cannot describe a physical reality either. Since (according to sec. 46) there is in any volume element of space a chance that a point event may happen, we can say, vaguely, that "matter is nowhere and everywhere". Certainly, we cannot say that atoms are small pieces of matter.

51. Operational Meaning of 'Matter'

A great part of this trouble comes from a misuse of the word 'matter'. In our everyday language we know very well what the operational meaning of 'matter' is. We know that iron or wood or the human body consist of "matter". But if we consider physical quantities which enter into the hypothetical setup of our physical science, we encounter words of which it is hard to say whether they mean "matter" or not. We use words like 'electricity', 'magnetism', 'ether', 'human mind', etc. The operational meaning of 'matter' is very clear if we speak of a piece of iron or wood or meat as being matter. Such a piece is identified as being "matter" by giving us the experience of resistance against penetration, of temperature, of color, of observable motion, etc. But if we ask whether we should call an electric charge or the ether "matter", we begin to doubt. For only a part of the operations which allow us to identify a piece of iron as matter are applicable in the case of an electric charge or of the ether. It is arbitrary to take just some particular operations as a characteristic of matter. The situation has been dramatized by the fight which quite a few philosophers have put up in order to keep in use the word 'matter' without any regard for its operational meaning. It has been suggested that we have to distinguish between the physical concept of matter and the philosophical one. While in physics "structure of matter" means the way our observable bodies are built up from protons, electrons, etc., we are told that the philosophical concept of matter should mean everything which exists objectively independent of our subjective sense experience.

In order to avoid any ambiguity and to keep strictly to the operational meaning, it seems that the most reasonable thing to do is to use the word 'matter' in the sense it is used in our everyday language. This means to call a table a piece of matter and our brain a piece of matter, but not to refer by the word 'matter' to concepts like electrons or photons, let alone the "ether" or the "mind".

52. Electromagnetic Mass

At the end of the nineteenth century it became clear that every electrically charged particle has a mass, m, which can be calculated from the charge and the size of the particle. The mass m is (sec. 11) defined by $f = ma$. If we have a spherical particle with the charge e (in electrostatic units) and the radius a, the mass is approximately e^2/ac^2, where c is the speed of light. This follows from the well-known law of the electromagnetic field, according to which an accelerated particle is equivalent to an electric current of increasing intensity. According to the law of self-induction, every current has a certain inertia against increase of its intensity. Therefore, an electrically charged particle is "recalcitrant" against an attempt to accelerate it. Therefore, it possesses an "inert mass" according to the operational definition of this term.

Occasionally one has distinguished between a "real mass" in the Newtonian sense of this word and an "apparent mass" which is feigned by an electric charge. However, these names can be misleading. They may suggest the metaphysical idea that an "apparent mass" is nothing "material" but something more subtle, something immaterial—electricity. It has been suggested on good grounds that every mass may be an apparent one and have its origin in an electric charge. For a neutral particle could consist of a positive and a negative charge each of which would have an apparent mass. They would not reveal their charges, which are neutralized. This electromagnetic theory of matter has been described by the slogan "matter has disappeared". It has been used largely in a crusade against materialism and on behalf of idealism. It is noteworthy that Lenin made this crusade the starting-point of his principal philosophical book. Actually the word 'matter' does not mean anything beyond the realm of the operational meaning mentioned above. The "hard fact" is that every particle carrying a charge e has, according to the operational definition of inert mass, a mass e^2/ac^2. Whether one has to introduce also a "mass" which is not covered by this formula is a question of the adjustment of our physical symbols to our experiences. It certainly has nothing to do with the question of the "reality" of matter.

53. The Number of Molecules in a Gas

To examine the structure of matter, it is obviously convenient to consider matter in the gaseous state. For, in this state, matter can be expanded arbitrarily and we can find out whether by this expansion we obtain the same type of matter in an infinitely diluted state or whether we obtain some small solid masses floating in a largely emptied space. If a gas is inclosed in a container, the phenomena taking place can be interpreted conveniently by the assumption that the gas consists of small particles which fly around until they hit the walls. The average kinetic energy is, according to the kinetic theory (sec. 19), proportional to the absolute temperature. These particles are called "molecules". They are certainly not pieces of "matter" in the sense in which we suggested in section 51 that this word be used. For such a particle has no temperature, no density, no color, but it has at every instant of time a momentum and a kinetic energy. It has certainly no surface as a piece of iron has, but it has a certain symmetry which may be spherical, cylindrical, etc. The salient point is whether the number of these molecules in one cubic centimeter can be figured out. What is the operational meaning of this number? If there are several independent operations by which it can be figured out, we would say, with P. W. Bridgman, that these molecules are a "physical reality". This number can indeed be figured out. It can be done, e.g., by the observation of the kinetic energy of particles of microscopic visibility floating around in the gas and pushed around by the molecules. From the statistical hypothesis it can be derived that such a particle has the same kinetic energy as a gas molecule. Since we know the total energy of all molecules in the container from the observation of the pressure exerted upon the walls, we can divide this energy by the energy of a single molecule and obtain the number we are looking for. One obtains L = approximately 6.8×10^{23} molecules in one mole (Avogadro's number).

Today there is even a method by which this number can be obtained in a much more direct way. In some cases one can spread out a substance in a layer which has the thickness of only one molecule. If we know the original volume of the substance and the area of the monomolecular layer, we can calculate the number of molecules. We obtain again the same number L. This number is the clue to the structure of matter. If we condense a gas, the molecules will touch each other approximately. We can obtain from the volume of the condensed state and the number L of the molecules the size of a molecule. We find that such a molecule is approximately

10^{-8} cm. across. This length—10^{-8} cm. —is a characteristic distance in the realm of molecules. It is called one "Ångstroem unit" (1 Å).

In a solid body, e.g., in a crystal, the molecules are almost in touch with each other. Therefore, the distance between two molecules in a solid body is also approximately equal to one Ångstroem. If the substance considered is not a chemical element but a compound, the molecule consists of a number of smaller particles called atoms. The table-salt molecule (sodium chloride, NaCl) consists of one atom of sodium and one atom of chlorine. The size of the atom is of the same order of magnitude as the size of a molecule. The molecule of a chemical element may consist of one atom, like mercury, or of several equal atoms, like hydrogen (H_2). The operational meaning of the number of molecules is very clear in a gas. In a container of the volume V at the temperature T, each molecule exerts the pressure KT/V on the wall, where K is a universal constant. In a solid body it is not unambiguous which atoms must be regarded as belonging to one and the same molecule. If we consider, e.g., a sodium chloride crystal, the sodium and chlorine atoms form a lattice within which there are no subdivisions. It is very arbitrary to pick just one sodium atom and one of the neighboring chlorine atoms and to call them a molecule. The statement that "one particular Na atom with one particular Cl atom forms a molecule" has no operational meaning.

54. The Problem of Chemical Binding

It has been an old problem to explain how the atoms in the chemical compound are kept in their position. A certain equilibrium between attraction and repulsion is needed. Is it possible to account for these actions on the basis of Newton's laws of motion and the laws of electromagnetic forces? The forces binding the atoms in the molecule are referred to as "valence forces". In the sodium chloride molecule the atoms are ionized. There is a positive and a negative electric charge. The attractive valence forces are electrostatic attractions between unlike charges. The repulsive forces are exerted by the shells of the atoms which are all negatively charged (secs. 58 and 59). But the matter is different if we have to derive the binding between the two hydrogen atoms in the hydrogen molecule. Between these neutral atoms there can be no attractive force in the Newtonian sense. The phenomenon of binding can be explained only on the basis of wave mechanics in which the concept of the path of a particle, and, in particular, of a particle at rest, has no operational meaning. This point will become clear when we discuss the structure of the atom (sec. 57).

55. The Structure of the Hydrogen Atom

We shall discuss first the simplest atom—the hydrogen atom—and describe later how the atoms of other elements differ from it. As already mentioned, the hydrogen atom, which has a size of approximately 10^{-8} cm. = 1Å, consists of two smaller particles, the proton and the electron. It has been accepted as a fundamental hypothesis in physics that every electrical charge is an integer multiple of the elementary charge $e = 4.8 \times 10^{-10}$ e.s.u. The proton and electron are each charged with one elementary charge. But the electron has a negative, and the proton a positive, charge. If we know the charge, we can determine the mass of a particle by simple experiments on the basis of Newton's laws of motion. If we denote the acceleration by a, the mass by m, the intensity of the electrostatic field in which the particle is moving by E, Newton's law of motion says that $ma = eE$ or $a = e/mE$. By observing a and E, we can determine e/m, and, from knowing e, we can calculate m. Since we have good reason to identify the cathode rays with flying electrons, the canal rays with flying protons, we find, by measuring the acceleration of these rays in electric and magnetic fields, that the mass of the electron is 9×10^{-28} gm., while the mass M of the proton is 1,837 times greater. These subatomic particles are, of course, of much smaller size than the atom.

We can estimate their size in different ways. If we assume, e.g., that the mass m of the electron can be derived entirely from its charge, we have $e^2/r = mc^2$, where r is a radius of the electron (sec. 52). If we substitute into this formula for e the charge and for m the mass of the electron, we find for the radius the value $r = $ approx. 10^{-13} cm. We call the proton the "nucleus" of the hydrogen atom because its mass M is much greater (about 1,800 times greater) than the mass of the electron. According to the original presentation of atomic physics, due to Rutherford, the electron traverses an orbit around the nucleus. The nucleus is of a size similar to that of the electron. We shall discuss later how the nuclear size can be figured out. This presentation kept strictly to the language of Newton's mechanics. In this pattern of description the space between electron and nucleus is empty. There is no "matter" within the atom, and one could use this argument in favor of the philosophy of "dynamism", according to which there is no matter at all in the world but only centers of forces. The nucleus attracts the electron according to Coulomb's law, $e'e/r^2$, when e' is the charge of the nucleus. If we calculate the smallest orbit of the electron around the nucleus, we find again that this radius is approximately $r = 10^{-8}$ cm. (sec. 39). We found the same value for the size of the atom starting from the number L of molecules (sec. 53).

If we remember our presentation of wave mechanics, we must understand that it is a very perfunctory way of describing the hydrogen atom to say that the hydrogen atom consists of two particles which are very small compared with the diameter of the atoms, the space between them being empty.

As a matter of fact, these statements have operational meaning only within the range of Newton's mechanics; since the radius of curvature of the "orbit" of an electron would be comparable with the De Broglie wavelength (sec. 40), Newton's concept of the "path of a particle" has no operational meaning for an electronic orbit near the nucleus. The real description of the hydrogen atom is given by a function which describes to us the chance of finding the electron at a certain point in space if we make an experiment which could reveal the presence of an electron at this point. Or, speaking more exactly: If we bring into the vicinity of the nucleus a measuring instrument which is able to register point events (e.g., scintillations on a screen), we find that along the circle with the radius $r = 1$ Å such an event happens much more frequently than at other points. However, the event may happen at any point in space which is near to the nucleus. To say, therefore, that the space within the atom is "empty" is to use the word 'empty' in a sense which has no operational meaning. For the distinction between the part of space which is penetrated by a certain orbit and the part of space which is empty is a distinction which has an operational meaning only within the realm of Newtonian mechanics. Even the space "outside" the atom cannot be called "empty" in the original sense of this word. For at any point in space there is a slight chance that "the electron may be caught"; speaking exactly,

that a "point event" can happen (sec. 46).

56. Electron, Proton, and Neutron

The mass of the hydrogen atom is equal to the sum of the mass of the proton plus the mass of the electron. This sum is not very different from the mass of the proton itself. The chemical reactions of hydrogen are determined by the fact that its nucleus has a charge $+e$ and its shell consists of one electron with the charge $-e$. It was discovered recently that there is an element which has the same chemical properties as hydrogen but has a nucleus with a double mass. This means that the new element, "heavy hydrogen", has the same nuclear charge $+e$ and the same shell as ordinary hydrogen. But its nucleus contains besides the proton also a second particle which has nearly the same mass as the proton but does not carry any electric charge. This particle which has played a great role in recent nuclear physics is called "neutron". Ordinary hydrogen and "heavy hydrogen" are the simplest examples of a couple of "isotopes", i.e., atoms which have one and the same nuclear charge but different nuclear masses. Speaking accurately, the mass of the neutron is a little larger than the mass of the proton. Perhaps the neutron itself consists of one proton and of one electron, for the charges of these particles would neutralize each other. If this were so, the neutron would be, in a certain way, a hydrogen atom on a smaller scale. The hydrogen atom consists of a proton and an electron at a distance of 10^{-8} *cm.* $= 1$ Å, but the neutron would consist of the same two particles but at a distance of only 10^{-12} cm. $= 10^{-4}$ Å from each other (nuclear size). As we have seen, the smallest orbit of an electron around the proton has a distance of 10^{-8} cm. according to wave mechanics as derived from Coulomb's law of attraction and Bohr's theory of the atom.

The nucleus of heavy hydrogen is also called "deuteron". Its existence and stability give clear evidence that orbits of such a small size (10^{-12} cm.) cannot be derived from Newton's and Coulomb's laws. For the constituents of a deuteron, the positively charged proton and the electrically neutral neutron, do not exert any force upon each other according to the laws of mechanics and electrostatics. The fact that they can form a coherent particle, the deuteron, can be derived from the new mechanics of small particles. If we assume that the neutron consists of a proton and an electron, the deuteron consists of two protons and one electron. If we examine the field of force produced by both protons according to traditional physics, the electron could be at rest only in the symmetry plane of both positive charges. However, according to wave mechanics, there is for any space element a certain fraction of the time of observation during which the electron is staying in this region. Therefore, it can also be near to either of the positive charges. As a matter of fact, it can easily be shown that the electron changes periodically its position between the two positive charges. 'Position of a particle' is to be interpreted in the complementary language of wave mechanics. Then each part of the system becomes alternately positive and negative. In this sense, the deuteron consists at any instant of time of two parts with different signs of charge. The sign of the charge on each side changes periodically. We call this way of attraction "exchange force", for this attraction is produced, vaguely speaking, from the fact that the electron changes its position among the two protons. These "exchange forces" are, of course, not "forces" in the strictly Newtonian sense. They determine no acceleration but only the frequency of a certain constellation of particles (speaking more exactly, of certain point events).

57. The Forces of Chemical Binding

The existence of the neutron and the deuteron throws some light upon the dark spots in the problem of valence forces in chemistry. As a neutron is on the nuclear scale of 10^{-12} cm. a reproduction of the hydrogen atom, the deuteron nucleus is a reproduction of another well-known particle—the ionized hydrogen molecule. The hydrogen molecule consists of two hydrogen atoms, i.e., of two protons and two electrons. If we remove one electron, we obtain a particle with one elementary charge $+e$, called a positive "hydrogen molecule ion". It consists obviously of two protons and one electron exactly as does the deuteron nucleus. But, while this nucleus has a size of 10^{-12} cm., the ion is of atomic size. The distance of the electron from the protons is approximately 10^{-8} cm. However, the existence of such a compound of two positive charges can be explained exactly as in the case of the neutron-proton compound by exchange forces. One can understand that the introduction of wave mechanics has contributed much toward the understanding of valence forces, the saturation of valences, and other problems of chemical binding (sec. 54).

58. The Structure of the Atoms of Different Chemical Elements

Two proton-neutron pairs form a new particle which has a charge $+2e$ and a mass of four proton masses. It is called a helium nucleus and is of nuclear size. If two electrons move around this nucleus at a distance of approximately 10^{-8} cm., we obtain a helium atom. The helium nucleus has been known for a long time as the alpha particle emitted by radioactive substances.

On the basis of this hypothesis about the structure of the simplest atoms, we can attack the ancient question about the structure of matter: Are the different chemical elements, iron, gold, sulphur, etc., really different substances, or are they only different configurations of one and the same substance? In physics this question has the following operational meaning: Is it possible to derive logically the chemical and physical properties of different elements like hydrogen, oxygen, iron, gold, etc., from the hypothesis that an atom of gold is

built up from exactly the same particles as an atom of oxygen, only in a different configuration? Previously the question was whether the atoms of gold, mercury, etc., are configurations of the simplest atom—the hydrogen atom. This seemed to be at least plausible. For the atomic mass of all elements is approximately an integer multiple of the atomic mass of hydrogen. Oxygen has an atomic mass which is approximately 16, nitrogen 14, and sulphur nearly 32 times the mass of the hydrogen atom. However, there are some serious exceptions. Chlorine, e.g., has the atomic mass 35.457. But exact measurements have shown that even the apparently integer numbers are only approximately integers. Nitrogen, e.g., is not exactly 14, but 14.008. Since we know that the atom itself is a compound of subatomic particles, at least of protons and neutrons, we understand that it was an oversimplification to expect all atoms to be just different configurations of hydrogen atoms.

We can expect only that all nuclei are compounds of neutrons and protons. Then, of course, the manifold of possible configurations is much greater, and we must not expect that the atomic masses are all integer multiples of the hydrogen mass. The generally accepted hypothesis is now that every atom has a nucleus which consists of N neutrons and P protons. They are packed so tightly together that they form a particle of nuclear size 10^{-12} cm. or 10^{-4} Å. The electric charge of such a nucleus is obviously

$+Pe$ and the mass approximately $P + N$ proton masses. This nucleus is surrounded at a distance of about 10^{-8} cm. by P negatively charged electrons. The total charge of the atom is zero. It is a neutral particle. Each nucleus is characterized by two integer numbers: the "atomic number" P and the "mass number" $A = P + N$. For the "ordinary hydrogen" $P = 1$, $A = 1$, for the "heavy hydrogen" $P = 1$, $A = 2$, for helium $P = 2$, $A = 4$, for "ordinary nitrogen" $P = 7$, $A = 14$, etc. As there is besides the "ordinary hydrogen" the isotope "heavy hydrogen" with the same P and a different A, there are besides the "ordinary nitrogen" several "isotopes of nitrogen" which have all the number $P = 7$ protons in common but have "atomic masses" A different from 14.

59. Chemical Properties and Atomic Weights

The chemical properties of an element, e.g., oxygen, depend upon the nuclear charge $+Pe$, which gives us also the number P of the electrons in the shell. Speaking exactly, the electrons are distributed among several shells. But the distance of the nearest shell from the nucleus is of the order of magnitude of 10^{-8} cm. Without going into details, we understand that chemical reactions take place in the following way: If two atoms (e.g., one sodium and one chlorine atom) approach each other, one electron of the outer shell of chlorine leaves its place and enters the outer shell of sodium. In this way these two atoms become of unlike electric charge (are ionized) and attract each other.

The chemical reactions of atoms are not affected by the number N of neutrons in the nucleus, provided that the number P of protons in the nucleus and electrons in the shell is not altered. We know already, as an example, the heavy hydrogen as distinct from the ordinary hydrogen. If atoms have equal nuclear charges Pe (i.e., equal atomic numbers P) but different numbers N of neutrons in the nucleus, they are "isotopes" and have different "mass numbers" A. Besides the ordinary oxygen $P = 8$, $A = 16$, there are isotopes with $P = 8$, but $A = 15$ and $A = 17$.

Two isotopes cannot be separated by chemical reactions. But as the masses of the nuclei are different, they can be separated by physical methods which make use of this difference, e.g., diffusion and centrifuging. If we have a piece of iron or a container filled with chlorine, we have obtained these samples by purifying them by chemical methods. But it can happen that our sample which is chemically pure contains not only atoms with one and the same number of neutrons but perhaps a mixture of isotopes which contain all the same number of electrons P which is characteristic of this chemical element. But the number of neutrons is different, and therefore the mass numbers are different too. If we have, e.g., a mixture of x atoms of ordinary hydrogen and y atoms of heavy hydrogen, the mass of the mixture would be $x + 2y = x(1+2y/x)$

proton masses. If y/x is, e.g., $1/10$, the mass of the mixture would be $1.2x$ proton masses. The number $1.2x$ is called the "atomic mass" of the mixture. It is the same number which is known from elementary chemistry as the "atomic weight" of an element like hydrogen. Actually it is the mass of a mixture of several isotopes of hydrogen. This "atomic mass" is certainly not an integer multiple of x. The departure of the atomic mass of elements from the integer multiples of hydrogen can be understood if we assume that, e.g., the gas which we call chlorine is actually a mixture of several isotopes of the ordinary chlorine. The departure from the integer value depends upon the composition y/x of this mixture.

However, even if we consider a chemical substance which is not a mixture of isotopes, the atomic mass cannot be expected to be an exact integer multiple of the proton mass (sec. 30). The "atomic mass" is not exactly equal to the "mass number". For example, in the case of helium ($A = 4$) we know that the nucleus consists of exactly four particles. If they were separated, their "atomic mass" would be "four". But by the formation of the helium nucleus potential energy decreases, kinetic energy is produced, and the rest mass must decrease (sec. 30). We know from the theory of relativity that the tight packing of the protons and neutrons in the nucleus must be responsible for a mass defect. The mass of the nucleus must be smaller than the sum of the masses of the pro-

tons and neutrons which constitute the nucleus. This fact accounts for the small departure of the atomic masses of chemical elements from the exact multiples of the proton mass, while the mixture of isotopes accounts for the "great" departures.

60. The Nuclear Forces

If we bombard atoms (e.g., a foil of gold) with alpha particles (helium nuclei with a charge $2e$), these particles are deflected by the repulsion of the positively charged gold nucleus. From the examination of the hyperbolic orbits of the particles under the influence of the deflecting forces we notice that this deflection is in some cases so strong that we have to assume that the alpha particles get very close to the center of the gold nucleus. Since the repulsive force is the Coulomb force (Pe^2/r^2), we can find from an easy calculation that the particle must approach the center of the gold nucleus to a distance which is smaller than 10^{-12} cm. One can also find from this calculation that for a distance greater than 10^{-12} cm. the Coulomb inverse-square law accounts well for the observed orbit. But it is equally certain that at a distance of 10^{-13} cm. from the center the inverse-square law is no longer valid. At a distance of 10^{-13} cm. a new type of force comes into play—the "nuclear force".

We have to do here not with a "force" in the Newtonian sense but rather with a kind of "exchange force". These "forces" are negligible at a distance which is considerably more than 10^{-13} cm. from the center. The radius of the electron can be calculated (sec. 52) as $a = e^2/mc^2 = 10^{-13}$ cm., if we assume that the mass of the electron has its origin only in its electric charge. If we assumed the same thing for the proton, its radius would be e^2/Mc^2, where M is the mass of the proton: $M = 1,837$ m. Therefore, the size of the proton would be only about 10^{-16} cm. The size of the nuclei would be much smaller than the size of the electrons. However, the examination of the deviation of alpha particles by the nuclei of different substances shows that the size of the nucleus cannot be so small. It must be, as a matter of fact, of the order of magnitude of 10^{-13} cm. This means it must be approximately the same as the size of the electron. This consideration shows that the proton mass cannot be completely derived from the electric charge; there must be forces other than the electromagnetic ones. These nuclear forces, which are analogous to the chemical valence forces, distinguish the phenomena in the nucleus from the phenomena at a great distance.

The chemical reactions and the emission of spectral lines (of visible light and X-rays) are phenomena in the shell or, exactly speaking, phenomena which can be predicted from what happens in the shell. However, there are other phenomena which have their foundations in the happenings within the nucleus.

61. The Phenomena of Radioactivity

The spontaneous emission of alpha particles by some atoms, in particular the radium atom, was one of the earliest discoveries which foreshadowed the twentieth-century physics. This emission is the first example of an atomic phenomenon which originates in the nucleus and cannot be understood on the basis of Newton's mechanics. According to section 46, it can be derived from wave mechanics. From this theory we can predict the percentage of alpha particles which leave the nucleus per second. But we cannot predict which specific particle will succeed in leaving the nucleus. It has been known for a long time that no physical law predicts which alpha particle will leave the nucleus but only what percentage of all particles present in the nucleus are emitted per second. This percentage accounts for the rate of decay which is characteristic for every radioactive substance.

The radioactive phenomena which were originally regarded as peculiar exceptions to the well-known laws of physics have now become a typical example of the application of wave mechanics. Originally one knew only an emission of alpha particles to be a spontaneous reaction within the nucleus of radioactive atoms which occurs without any interference from without. But later it was discovered that disintegration of nuclei can be produced artificially if we bombard nuclei with protons, neutrons, or alpha particles. The "stable" nuclei can be disintegrated by bombardment, while the unstable (radioactive) ones disintegrate spontaneously. It was also discovered that by bombardment a stable nucleus can be converted into an unstable one (artificial radioactivity).

62. Production of Positrons, Electrons, and Photons by Nuclear Reactions

As a result of bombardment experiments not only the particles which have been known before can emerge but a new particle was discovered, the "positron", which has the positive electronic charge $+e$ and a mass which is as small as the electron mass.

Whereas protons and neutrons are believed to be present in the nuclei of all chemical elements, the electrons, positrons, and photons which are emitted by the nucleus during a great many reactions have not been parts of the nucleus before but are produced by the reaction. We know that the atom emits light (photons) when its energy drops, and nonetheless these photons have not been present in the atom before. In the same way the nucleus emits photons which are produced by the transition from one state of the nucleus into another state with a lower energy. The photons emitted by the nucleus are, however, of a much higher frequency than the light emitted by the atoms. The "nuclear photons" are the "gamma rays" which are known from radioactive radiation.

These considerations make it very clear that the nuclear reactions cannot be described by using the language of the mechanics of everyday life. We cannot simply say that the beta particles or the alpha particles are emitted from the nucleus as bullets are shot from a gun. Even the removal of a particle by bombardment cannot be described in this language of the mechanics of medium-sized bodies. According to Niels Bohr, a particle (e.g., a neutron) which strikes the nucleus sticks first in the nucleus like a bullet fired into a box of sand. Then a reaction takes place between the intruding particles and the nucleus. By this reaction new particles are produced. The intruding particle does not simply throw out other particles as a ruffian intruding in a crowded streetcar would throw out other passengers. The emission of a beta particle (electron) from the nucleus follows laws which are more involved. If the nucleus experiences a certain energy drop ΔE, a photon (gamma ray) of a definite frequency ν is emitted. ν is determined by $h\nu = \Delta E$ (sec. 39). The beta particles which are emitted have a continuum of kinetic energies.

63. The Structure of the Nucleus

If we try to make a picture of the structure of the nucleus, we can do it at least in a perfunctory way by using the model of a "drop of fluid" suggested by Bohr. The forces by which the nucleus is kept together do not belong to the electromagnetic type but to the exchange forces which account for a great part of chemical binding between the atoms in a molecule and for the cohesion between the atoms and molecules in a piece of matter.

These nuclear forces produce an effect which can be compared to the surface tension which is responsible for the cohesion of a drop of fluid. On the other hand, the positively charged protons exert repulsive forces upon each other. A stable nucleus can exist only if there is equilibrium between this "surface tension" which is proportional to the area of the surface and the repulsive forces of the protons.

If we consider a nucleus with few protons and neutrons, we notice that the number of protons and neutrons are almost equal. For light nuclei we find approximately $P = N$ or $A = 2P$. For oxygen, e.g., we know that $P = 8$. The atomic mass A is for "ordinary oxygen" exactly $A = 16$. There are, however, isotopes with $A = 17$ and $A = 18$, where $A = 2P$ is only approximately fulfilled. We can conclude, therefore, that for these light nuclei surface tension and electrostatic repulsion are in equilibrium if the number of neutrons equals approximately the number of protons. But, if we consider heavy nuclei, the area of their surface by unit of volume becomes smaller. Therefore, the number of neutrons which are needed to produce a sufficient surface tension becomes greater. In the case of the heaviest nucleus ($P = 92$) there are 146 neutrons in the "ordinary uranium", while for a rare isotope we find $N = 143$. This means that the mass numbers are $A = 238$ and $A = 235$, which are both considerably greater than $2P = 184$.

It is obvious, therefore, that, by splitting a heavy nucleus, such as uranium, into two lighter nuclei, some of the neutrons are liberated, as they are not needed to secure the stability of the lighter nuclei. There is no stable nucleus with $P = 90$ or greater.

As we know from the theory of relativity, the "binding energy" which keeps the particles together is connected with the "mass defect" of the nucleus (sec. 30). If we compare the sum of the masses of the particles (protons and neutrons) with the measured atomic weight of a chemical element, we find a certain difference which is due to the energy which has been used in the formation of the nucleus. If we assume that we have to do not with a mixture of isotopes but with a sample which consists of identical nuclei of the atomic number P and the mass number A, the atomic weight W which is measured would be $W = A = P + N$, if we computed the atomic weight by adding up the masses of the particles from which the nucleus is built (neglecting the small difference between proton mass M and neutron mass). As a matter of fact, the measured atomic weight W is a little smaller than the mass number A. If this mass defect is $(A - W)M$, the "binding energy" of the nucleus B is $B = c^2M(A - W)$, which has to be supplied in order to break up the nucleus into its elementary particles. This energy is also liberated by the formation of the nucleus.

It is particularly important to know the binding energy of a nucleus divided by the number of particles of which this nucleus is built. This number B/A has, in first approximation, one and the same value for all nuclei. This fact checks well with the picture of the nucleus as a drop of fluid. For in a fluid the energy is proportional to the volume, and the volume of a nucleus is proportional to the number A of particles.

If we examine the values of B/A more closely, we notice a clear trend. The "binding energy per particle" is increasing from the lightest nuclei to the value $A = 60$ (nickel) and is again decreasing toward the heavy nuclei. The nuclei in the middle of the series are the most strongly bound.

If we speak of a nucleus being comparable to a drop of fluid, we do not mean, of course, that the behavior of a nucleus can be predicted exactly by the equations of fluid motions as treated in ordinary hydrodynamics. For this would mean that the particles in the nucleus obey Newton's laws, which is outruled by the experiments on the behavior of small bodies (Part VII). The similarity to a fluid is restricted to the validity of some algebraical and dimensional relations which have the same form as the relations between surface tension and repulsive forces in a fluid. It is noteworthy, however, that the behavior of these extremely small particles can be predicted, as far as some general features are concerned, by this picture. It would, however, be a great mistake if anyone should take this picture literally and draw conclusions from Newton's laws in an attempt to show that the motions of fluid particles contradict the behavior of smaller particles as predicted by wave mechanics. One must never forget that the term 'particle' in the description of the nucleus has not the same operational meaning as this term has in ordinary mechanics. 'Particle' in nuclear physics is only an abbreviation which refers to "point events" which can be statistically predicted.

64. Atomic Power[28]

We learned from section 63 that the formation or the breaking-up of a nucleus can become a source of energy (atomic power). We consider not only the formation of a nucleus from the "elementary particles", protons and neutrons, but from any smaller nuclei. If the binding energy of the resulting nucleus is smaller than the sum of the binding energies of the "building stones", the formation produces energy and rest mass disappears. If, however, the binding energy of the resulting nucleus is greater than the energy sum of the constituents, energy is liberated in breaking up this nucleus, and the fragments together have a smaller mass than the original nucleus.

If A increases, the binding energy per particle, B/A, decreases among the lighter nuclei and increases among the heavier nuclei. By this fact two ways are suggested for the production of energy from nuclear reactions. One could either use the formation of a light nucleus or the disintegration of a heavy nucleus.

The first way was used by nature in our sun, the second one by human inventors in the "atomic bomb". According to a widely accepted astrophysical theory, the heat of our sun is reproduced by a cycle of nuclear reactions (Bethe's Carbon Cycle) which boils down eventually to the formation of helium nuclei from protons.

In the first atomic bomb an isotope of uranium ($P = 92$, $A = 235$) was used. This heavy nucleus can be broken up by neutron bombardment. Every uranium nucleus of 235 particles is split into two nuclei of nearly equal size and some free neutrons. As the binding energy of all the fragments together is much smaller than the energy of the original uranium nucleus, there is a considerable production of kinetic and radiant energy accompanied by a disappearance of rest mass. The liberated neutrons hit other uranium ($A = 235$) nuclei and produce a further disintegration. In this way a "chain reaction" is started, which produces in a short time a considerable amount of energy in a way similar to that in which a lighted match can start a forest fire, by a "self-sustained" reaction.

This kind of splitting into two particles of similar size has been known under the name of "uranium fission". In order to start this chain reaction, one does not need to produce neutron projectiles first. There are always some neutrons in the air, originating from cosmic rays or radioactive substances. According to wave mechanics (sec. 46), there is even a certain chance that fission starts spontaneously. By the disintegration of 1 pound of uranium (235) an energy of nearly eleven million kilowatt hours is produced.

The process which takes place in the atomic bomb is a good example to offer in showing that disintegration of nuclei is a very common phenomenon which is not restricted to the radioactive processes in the older sense (emission of alpha and beta particles).

Besides the uranium bomb, a second type has been used which provides a good example of fission combined with artificial radioactivity. If ordinary uranium ($P = 92$, $A = 238$) is bombarded with neutrons, sometimes a neutron is captured by a nucleus. When this happens, a radioactive emission (of beta particles) takes place. This can be interpreted by assuming that one neutron in the nucleus emits an electron and is converted into a proton (sec. 62). The number of protons P now becomes 93, and we have a new element "neptunium". By again capturing a neutron, and again emitting an electron, the neptunium becomes "plutonium", with $P = 94$. This new element is the material of the second atomic bomb. It suffers fission like uranium ($P = 235$) and can start a chain reaction.

The explosion of the bomb has proved that Einstein's law $E = m_0 c^2$ (sec. 30) plays a role not only in subtle laboratory experiments but also in power engineering. The introduction of the speed of light c into the equation of mechanics, which has been thought of as a far-fetched sophistication, becomes now conspicuous by the enormous energy production which can be understood only by the enormous value of c^2, the square of light velocity.

The explosion of the atomic bomb provided a very conspicuous example of the possibility that "mass can disappear". As a matter of fact, the disintegration of one pound of uranium ($P = 235$) left only 0.999 of a pound of fragments, while 0.001 of a pound was "converted into energy" (radiant and kinetic). We must not forget, however, that it would be misleading to describe this fact by saying that "matter has disappeared". We can say this if we mean "rest mass" by the word 'matter'. But if we gave to the sum of rest mass and E/c^2 the name 'matter', we could say again that "matter was conserved" even through the atomic explosion. If we use the term 'matter' beyond its common-sense meaning (sec. 51), we must give it an arbitrary operational definition. As in the case of 'time' or 'length', some operations which are to measure the "quantity of matter" yield identical results, if

we use the word 'matter' in the traditional sense and assume that the laws of traditional physics are valid. But if the laws of physics were different, it might happen that these operations would render different results, and it becomes ambiguous which of them is the "real quantity of matter". An example of such operations is provided by the measuring of masses before and after an atomic explosion. If the relativistic mechanics is true, the "masses" before and after the explosions are not equal. Therefore, "matter" has disappeared, if we identify "matter" with "rest mass". But if we choose a different operational definition of matter, we could not say that "matter has disappeared".

This statement has only an unambiguous operational meaning if we stick to traditional physics and its terminology. But in the new physics we cannot say whether "matter" does disappear or not, because the word 'matter' can no longer be used unambiguously, without introducing a new operational definition. As we learned in section 2, the operational meaning of an expression can only be defined unambiguously if we assume the validity of certain physical laws.

IX. Conclusion

It is hardly possible to decide accurately whether a certain presentation of physics is fit to become a part of a unified presentation of the sciences. However, it is easy to ascertain

that quite a few presentations of physics have decidedly produced confusion when one has tried to fit them into the structure of other fields of knowledge.

In biology, for instance, the physics of the twentieth century has been used to bring about a decision in the age-old conflict between the "mechanistic" and the "vitalistic or organismic" conception. We have been told that the "new physics" decides in favor of spontaneity and emergent evolution and against rigid mechanical explanations.

In medicine the "new physics" has been quoted as favoring the case of all sorts of "practitioners" in their old fight against "orthodox school medicine". For, by the principle of uncertainty, a certain flexible element has been brought into science which, according to these people, eases the rigidity of scientific rules and provides a place for methods which are based more on intuition than on systematic knowledge.

In sociology and economics the new physics has been employed to bolster the "organismic view" of human society sponsored by some religious and political groups against the "mechanistic view of society" which has been allegedly the view of liberal and Marxist economists.

In theology the new physics has been put into the service of the fight of religion against materialism by pointing out that now "freedom of will" has won a certain place in the formerly rigidly deterministic universe.

Even the "occult sciences" have seen a green light in the physics of the twentieth century. The theory of relativity introduced the conversion of matter into energy and has supported the belief in "dematerialization".

As we learned in Part II, every physical theory consists of three kinds of statement: equations between physical quantities (relations between symbols), logical rules, and semantical rules (operational definitions).

The sensational results of the application of the "new physics" to biology, sociology, medicine, etc., have been very frequently achieved by the following method: The symbols of physics (e.g., waves and particles) have been inserted into the sciences of biology, sociology, medicine, etc., but the operational definitions of the symbols have been omitted. This means that the physical theories have been applied in a crippled condition. As symbols without operational meaning do not lead to any palpable result, some kind of operational definition had to be introduced. Actually, those symbols have been interpreted as meaning what they traditionally mean in biology or sociology. This meaning is, of course, very different from their operational meaning in physics.

We see that the application of the "new physics" to other fields of knowledge has not always been made by fitting the physical theories in their scientific form into the structure of other sciences. No wonder that nothing but confusion could be the outcome of such an attempt to "integrate human knowledge".

Examples are obvious. If we say in physics that in the realm of the subatomic phenomena the concepts of mechanics cannot be applied, we mean to say that the symbols of classical mechanics, like 'position of a particle', have no operational meaning in the realm of the subatomic phenomena. However, in biology there has been an ancient conflict between the mechanistic and the vitalistic conception. To say that the "conceptions of mechanics cannot be applied to the phenomena of life" has always meant that life is something autonomous, spontaneous, emergent, etc. The symbols of the new subatomic physics, if understood in connection with their original operational meaning, allow a description of the subatomic phenomena which has nothing to do with spontaneity, emergent evolution, or purposiveness. This "new physics" is, in the language of biology, not less "mechanistic" than classical physics. Therefore, if we understand all statements of physics in their operational meaning, we cannot draw any conclusion in favor of a vitalistic biology.

If the statement "length is relative" is understood in its operational meaning, it is a statement about the fact that certain procedures of measurement yield different results, whereas it had formerly been believed that they yield identical results. But if we transplant the statement "length is relative" without its operational meaning into psychology or sociology or medicine, the word 'relative' is interpreted, of course, in the way in which this word has been traditionally used in these fields of knowledge. It has meant there that all knowledge is subjective or his-

torically and ethnically conditioned. If we carried the original operational meaning of "relativity" from physics into psychology or medicine, we could never arrive at a statement about subjectivity or vagueness, let alone "agnosticism".

What we can learn from these examples is simply the fact that a presentation of physics is fit to be a part of the unified sciences only if the operational meaning of every statement is explicitly formulated and carefully carried along when the statement is applied to other sciences.

While a physicist works within his own domain, the operational meaning of his statements is "understood". No explicit logical analysis is needed, and such an analysis would even be tiresome and pedantic. When, however, a statement of physics is transplanted into a foreign soil, the operational meaning is no longer "understood", and a careful logical analysis is indispensable. Otherwise, we have to approve statements like: "The first explosion of an atomic bomb exploded once forever the doctrine of the materialists" or "was a crucial experiment in favor of dematerialisation" or "spiritualism is confirmed by science".

While one lives within a limited group of people, it is "understood" what we mean by calling a man a "nice fellow"; but if one comes in contact with an unknown group, there is not even a guess as to whether by calling a man a "nice fellow" we hint that he goes all-out for prohibition or that he likes a good drink.

In a presentation of the "new physics" we must, therefore, distinguish very carefully between statements which ascribe to words or symbols a new operational meaning and statements which describe the results of new experiments.

But we must not fail, on the other hand, to understand the close interconnection between these two kinds of statement. As we learned in Part II, and confirmed by the examples of Parts V (relativity) and VII (wave mechanics), symbols and expressions are introduced into physics only if they are of some help in formulating physical laws—the laws predicting observable phenomena. This means that physical operations can serve as definitions of physical quantities only if a great many operations lead to an identical numerical result. "In the discovery of what operations may be usefully employed in describing nature is buried almost all physical experience. . . . . The discovery that the number obtained by counting the number of times a stick may be applied to an object can be simply used in describing natural phenomena was one of the most important discoveries ever made by man".[29]

As we learned in the Introduction, quite a few physicists have suggested that the misuse of physics in other fields of knowledge could be prevented by isolating physics from any contact with people who want to interpret its meaning for human thought in general. By such an ostrich policy physics would be converted into a collection of rules which might be useful in technology. But physical science would lose its traditional role as the vanguard of progressive thought.

A danger of this kind has arisen again and again in the history of thought. In the year 1907 the French philosopher and historian of science Abel Rey wrote: "If physical science which has had an essentially emancipating effect in history goes down in a crisis which leaves it only with the significance of a technically useful collection but robs it of every value in connection with the cognition of nature, this must bring about a complete revolution in the logical art. The emancipation of the mind, as we owe it to physics, is a most erroneous idea. . . . . One must restore to a mystical sense of reality everything that we believed had been taken away from it".[30]

This danger cannot be met by isolationism in physics. The only thing to do is to make "physics safe for its role as a part of the unified sciences". This means again to give careful thought to the logical analysis of physics and to be always aware that only the combination of relations between symbols, logical rules, and operational definitions constitute the science of physics. There are quite a few philosophers who have felt sorry that "behaviorism" and "logical empiricism" have banned words like 'soul' and 'mind' from scientific psychology. They may find comfort in the fact that the same schools of thought have banned words like 'matter' from scientific physics. A "soulless" psychology and a "matterless" physics have been established as parts of "Unified Science". Words like 'matter' and 'mind' are left to the language of every-day life where they have their legitimate place and are understood by the famous "man in the street" unambiguously.[31]

Notes

1. "I have not attempted to introduce a new philosophy into science, but rather to remove an obsolete philosophy which has occasionally survived in the writings of scientists for a longer time than in the writings of the philosophers themselves" (Ernst Mach, *Erkenntnis und Irrtum* [1905]). "To neglect philosophy when engaged in the re-formation of ideas is to assume the absolute correctness of the chance philosophic prejudices imbibed from a nurse or a schoolmaster or current modes of expression" (A. N. Whitehead, *The Principle of Relativity* [1922]).

2. Philipp Frank, "Why Do Scientists and Philosophers So Often Disagree?" *Review of Modern Physics*, Vol. XIII (1941), and "The Philosophical Meaning of the Copernican Revolution," *Proceedings of the American Philosophical Society*, Vol. CLXXXVII (1944).

3. W. A. Wick, *Metaphysics and the New Logic* (1942); A. C. Benjamin, "The Unholy Alliance between Operationalism and Positivism," *Journal of Philosophy*, Vol. XXXIX (1942).

4. This *Encyclopedia*, Vol. I, No. 3, § 23 (Carnap).

5. *Ibid.*, § 8.

6. *Ibid.*, § 4.

7. *Ibid.*, § 24.

8. P. W. Bridgman, *Logic of Modern Physics* (1927).

9. It has been suggested by some authors that some rules of traditional logic be altered in order to fit modern quantum mechanics into a formalized system: M. Strauss, in *Erkenntnis*, Vol. VI (1936); L. Rougier, "La Relativité de la logique," *Journal of Unified Science*, Vol. VIII (1936); and G. Birkhoff and J. v. Neumann, "The Logic of Quantum Mechanics," *Annals of Mathematics*, Vol. XXXVII (1936).

10. Bertrand Russell, *The Scientific Outlook* (1931).

11. P. W. Bridgman, *The Nature of Thermodynamics* (1941).

12. Conventionalism has been the starting-point of a great many attempts to justify metaphysical creeds by arguing that "science" cannot tell us anything about physical reality and that therefore a vacuum is produced within which extrascientific methods may operate at will (see E. Le Roy, *Rev. de met. et de mor.*, Vols. XVII and XVIII [1899, 1900]; Pierre Duhem, "Physique de Croyant," *Annales de phil. chrét.* [1905/6]). The scientific angle is stressed by E. Nagel, "Nature and Convention," *Journal of Philosophy*, Vol. XXVI ((1929).

13. Henri Poincaré, *Science et hypothèse* (1903); P. Duhem, *La Theorie physique, son objet et sa structure* (1926); Philipp Frank, *Le Principe de causalité et ses limites* (1937), p. 184; and P. W. Bridgman, *The Nature of Thermodynamics*.

14. P. Frank, *Le Principe de causalité et ses limites*, p. 73.

15. P. Frank, *La Fin de la physique mécaniste* (1936).

16. P. Frank, "The Mechanical vs. the Mathematical Conception of Nature," *Philosophy of Science*, Vol. IV (1937).

17. P. Frank, "Modern Physics and Common Sense," *Scripta mathematica* (1939).

18. P. Frank, "Relativity and Its Astronomical Applications," *Sky and Telescope*, 1942, p. 9.

19. R. v. Mises, "Die Krise der Mechanik," *Proc. Congr. f. Appl. Mech.* (Delft), 1924.

20. *Logic of Modern Physics*, p. 59.

21. "It is the recoverability of the original situation that is important, not the detailed reversal of the steps which led to the original departure from the initial situation" (Bridgman, *Nature of Thermodynamics*, p. 122). In a great many writings this distinction has not been made and the word 'reversibility' is used for 'recoverability'.

22. Lord Kelvin (1852) cautiously assumed the validity of the Second Law for "inanimate" systems only. For a discussion of the apparent conflict between organic evolution and increase of entropy see Lecomte du Nouy, *Biological Time* (1936), and Bridgman, *Nature of Thermodynamics*, pp. 208 ff.

23. Bridgman, *Nature of Thermodynamics*, pp. 148 ff.

24. Quite a few philosophers suggested the introduction of a "philosophical time concept", without a clear-cut operational meaning, in addition to the time concept based on physical operations as used in the theory of relativity: H. Bergson, *Durée et simultanéité* (1922); A. O. Lovejoy, "The Time-retarding Journey," *Philosophical Review* (1931). A time concept based on biological operations was introduced by Lecomte du Nouy, *op. cit.*

25. "Can Quantum-Mechanical Description of Physical Reality Be Considered Complete?" *Physical Review*, Vol. XLVIII (1935).

26. In Newtonian mechanics the operational meaning of 'force' is based on the measurement of the acceleration of a particle. In the mechanics of subatomic particles the operational meaning of 'force' can be reduced to the statistical distribution of point-events or impulse-events in space at a certain time.

27. Albert Einstein, "Physics and Reality,' *Franklin Institute Journal*, Vol. CCXXI (1936).

28. H. D. Smyth, *Atomic Energy for Military Purposes* (Princeton, 1945).

29. P. W. Bridgman, *The Logic of Modern Physics*, p. 27.

30. *La théorie de la physique chez les physiciens contemporains* (Paris, 1907).

31. "Philosophy deals with positive truth, yet contents itself with observations such as come within the range of every man's normal experience" (C. S. Pierce, *Collected Works*, Vol. I, sec. 241).

Selected Bibliography

I. RECENT WORKS ON THE FOUNDATIONS OF PHYSICS

BERGMANN, G. "Outline of an Empiricist Philosophy of Physics," *American Journal of Physics*, XI (1943), 248–58, 335–42.

BOHR, NIELS. *Atomic Theory and the Description of Nature.* New York, 1934.

BORN, M. *The Restless Universe.* London & Glasgow, 1935.

BRIDGMAN, P. W. *The Logic of Modern Physics.* New York, 1929.

———. *The Nature of Physical Theory.* Princeton, 1936.

———. *The Nature of Thermodynamics.* Cambridge, Mass., 1941.

BROGLIE, L. DE. *Matter and Light.* New York, 1939.

DARWIN, C. G. *The New Concept of Matter.* London, 1931.

DAVIS, H. T. *Philosophy and Modern Science.* Bloomington, Ind., 1931.

EDDINGTON, A. S. *Space, Time, and Gravitation.* Cambridge, 1929.

EINSTEIN, A., and INFELD, L. *The Evolution of Physics.* New York, 1938.

ELDRIDGE, J. A. *The Physical Basis of Things.* New York, 1934.

FRANK, PHILIPP. *Le Principe de causalité a ses limites.* Paris, 1937.

———. *Between Physics and Philosophy.* Cambridge, Mass., 1941.

LENZEN, V. F. *The Nature of Physical Theory.* New York, 1931.

LINDSAY, R. B., and MARGENAU, H. *Foundations of Physics.* New York, 1936.

MISES, R. VON. *Kleines Lehrbuch des Positivismus.* The Hague, 1939.

REICHENBACH, HANS. *Philosophical Foundations of Quantum Mechanics.* Berkeley, 1944.

RUSSELL, BERTRAND. *Mysticism and Logic.* New York, 1929.

———. *The Scientific Outlook.* New York, 1931.

STEBBING, SUSAN. *Philosophy and the Physicists.* London, 1937.

SWANN, W. F. G. *The Architecture of the Universe.* New York, 1934.

WATSON, W. H. *On Understanding Physics.* Cambridge, 1938.

Selected Bibliography

II. CLASSICS ON THE FOUNDATIONS OF PHYSICS

DUHEM, PIERRE. *La Théorie physique: son objet et sa structure.* Paris, 1906.

ENRIQUES, F. *Problems of Science.* Chicago and London, 1914.

MACH, ERNST. *Science of Mechanics.* Chicago and London, 1919.

POINCARÉ, HENRI. *The Foundations of Science: Science and Hypothesis, The Value of Science,* and *Science and Method.* New York, 1919.

III. PHILOSOPHICAL INTERPRETATIONS OF PHYSICS

BRAHMA, N. K. *Causality and Science.* London, 1939.

CASSIRER, E. *Einstein's Theory of Relativity.* Chicago and London, 1923.

———. *Determinismus und Indeterminismus in der modernen Physik.* Goteborg, 1936.

DINGLE, H. *Through Science to Philosophy.* London, 1937.

EDDINGTON, A. *The Nature of the Physical World.* New York, 1928.

———. *Philosophy of Physical Science.* New York, 1939.

HALDANE, J. B. S. *Marxist Philosophy and Science.* London, 1938.

JEANS, J. *The Mysterious Universe.* New York, 1931.

———. *Physics and Philosophy.* New York, 1943.

MACKAYE, J. *The Dynamic Universe.* New York, 1931.

MARITAIN, J. *La Philosophie de la nature.* Paris, 1936.

———. *The Degrees of Knowledge.* New York, 1938.

MONTAGUE, W. P. *The Ways of Things.* New York, 1940.

NORTHROP, F. S. C. *Science and First Principles.* New York, 1931.

REISER, O. *Philosophy and the Concepts of Modern Science.* New York, 1935.

WHITEHEAD, A. N. *The Concept of Nature.* Cambridge, Mass., 1920.

———. *Science and the Modern World.* New York, 1926.

DATE DUE

APR 26 '83			
FE 11 '85			
MR 2 '86			